Diffusion of Biomass Energy Technologies in Developing Countries

Second Edition

Board on Science and Technology for International Development

Office of International Affairs

National Research Council

NATIONAL ACADEMY PRESS
Washington, D.C. 1984

NOTICE: The project that is the subject of this report was approved by the Governing Board of the National Research Council, whose members are drawn from the councils of the National Academy of Sciences, the National Academy of Engineering, and the Institute of Medicine. The members of the committee responsible for the report were chosen for their special competences and with regard for appropriate balance.

This report has been reviewed by a group other than the authors according to procedures approved by a Report Review Committee consisting of members of the National Academy of Sciences, the National Academy of Engineering, and the Institute of Medicine.

The National Research Council was established by the National Academy of Sciences in 1916 to associate the broad community of science and technology with the Academy's purposes of furthering knowledge and of advising the federal government. The Council operates in accordance with general policies determined by the Academy under the authority of its congressional charter of 1863, which establishes the Academy as a private, nonprofit, self-governing membership corporation. The Council has become the principal operating agency of both the National Academy of Sciences and the National Academy of Engineering in the conduct of their services to the government, the public, and the scientific and engineering communities. It is administered jointly by both Academies and the Institute of Medicine. The National Academy of Engineering and the Institute of Medicine were established in 1964 and 1970, respectively, under the charter of the National Academy of Sciences.

This study has been prepared by the Board on Science and Technology for International Development, Office of International Affairs, National Research Council, for the Rockefeller Foundation, under Grant No. RF 80040.

Second edition, January 1984

International Standard Book Number 0-309-03442-6

Library of Congress Catalog Number: 82-82005

Printed in the United States of America

PANEL ON RENEWABLE ENERGY TECHNOLOGY DIFFUSION IN DEVELOPING COUNTRIES

Raymond C. Loehr, Cornell University, Ithaca, New York, USA, Cochairman

Alejandro Melchor, Asian Development Bank, Manila, The Philippines, Cochairman

J. E. M. Arnold, Food and Agriculture Organization of the United Nations, Rome, Italy

Andrew Barnett, University of Sussex, Brighton, England

Elizabeth Colson, University of California, Berkeley, California, USA

Sylvia Forman, University of Massachusetts, Amherst, Massachusetts, USA

Ricardo Giesecke, United Nations, Managua, Nicaragua

M. N. Islam, University of Engineering and Technology, Dacca, Bangladesh

Jacqueline Ki-Zerbo, Permanent Interstate Committee for the Struggle Against Drought in the Sahel, Ouagadougou, Upper Volta

A. K. N. Reddy, Indian Institute of Science, Bangalore, India

Roger Revelle, University of California, La Jolla, California, USA

Wu Wen, Guangzhou Institute of Energy Conversion, Guangzhou, People's Republic of China

CONSULTANTS

Peter B. Hammond, University of California, Los Angeles, California, USA

C. Anthony Pryor, Center for Integrative Development, New York, New York, USA

R. Bhatia, Harvard University, Cambridge, Massachusetts, USA

Russell deLucia, Meta Systems, Inc., Cambridge, Massachusetts, USA

Peter Hayes, University of California, Berkeley, California, USA

Raymond Noronha, The World Bank, Washington, D.C., USA

T. B. S. Prakasam, Metropolitan Sanitary District of Greater Chicago, Cicero, Illinois, USA

Everett M. Rogers, Stanford University, Stanford, California, USA

Vaclav Smil, University of Manitoba, Winnipeg, Manitoba, Canada

Timothy Wood, Sahel Regional Technical Coordinator for Fuel Conserving Technologies, Ouagadougou, Upper Volta

STUDY STAFF, BOARD ON SCIENCE AND TECHNOLOGY FOR INTERNATIONAL DEVELOPMENT

M. G. C. McDonald Dow, Associate Director/Studies

E. Griffin Shay, Professional Associate

F. R. Ruskin, Editor

Sherry Snyder, Reports Editor

Carol Corillon, Research Associate

Jeffrey Gritzner, Professional Associate

Noel Vietmeyer, Professional Associate

Hertha Hanu, Administrative Secretary

Irene Martinez, Text Processing Coordinator

Wendy White, Information Services Manager

PREFACE

This study reviews the factors--technical, economic, cultural, and political--that affect the diffusion of biomass-based energy technologies in developing countries, identifies opportunities and limitations, and makes recommendations for improving technology diffusion for meeting the energy needs of the rural and urban poor.

The report has been produced under the sponsorship of the International Relations Division of the Rockefeller Foundation. The study was carried out by an international panel, assisted by National Research Council (NRC) staff, a team of consultants, and by many knowledgeable and concerned people around the world. The NRC wishes to express its appreciation to all who contributed to this project and to the Rockefeller Foundation for its generous support.

CONTENTS

OVERVIEW

This report is concerned with the factors that influence the introduction and diffusion of selected biomass-based renewable energy technologies in developing countries. It is based on discussions with those involved in planning and implementing energy projects in developed and developing nations, and on published information. It is also based on visits by panelists, consultants, and National Research Council (NRC) staff to seventeen developing countries in the course of this study to observe renewable energy projects first hand. These countries are: Brazil, Colombia, Dominican Republic, Ethiopia, Fiji, Honduras, India, Indonesia, Jamaica, Mauritania, Papua New Guinea, People's Republic of China, Philippines, Sri Lanka, Tanzania, Thailand, and Upper Volta. Selected observations based on these visits are incorporated into this report.

Rapidly rising oil prices have resulted in rapid adjustment of consumption practices in the industrialized nations and in the affluent sectors of developing countries. Conservation and substitution have been accomplished on a scale that many thought impossible five years ago, and as a result, oil prices are temporarily stabilizing. For developing countries, the effect of rising oil prices was catastrophic in two ways: (1) their precarious economies were overwhelmed by the costs of their modest oil needs to the extent that most are deep in debt to the oil suppliers, and investment in industrialization and agricultural development has virtually disappeared; and (2) substitution of biomass for petroleum fuels in cooking and small industry has increased firewood and charcoal consumption in the urban areas to the point where meeting this demand is causing serious deforestation and erosion in the countryside.

At the same time, renewable energy technologies are not being adequately diffused among the people most dependent on them--the rural and urban poor. Governments seldom recognize the importance of these technologies, and investment to support them is not attracted because the benefits are delayed. The poor are understandably reluctant to adopt technologies that do not meet their perceived needs, and they are seldom consulted when projects are being designed. Unless this situation is remedied, they will continue to be faced with serious shortfalls in food production and indefinite dependence on outside assistance. On the other hand, renewable technologies can be successfully adopted by poor people when they are involved in the planning and management of the project, and when, moreover, these projects can be profitable.

The technologies discussed include the generation of biomass through fuelwood plantations and agroforestry and the use of biomass in improved cooking stoves, charcoal manufacture, thermal gasification, and the production of biogas and fuel alcohol. These were selected because of their relevance to agricultural productivity and the dependence of the poor on biomass as an energy source.

For each of these technologies, the technical, economic, social, and cultural factors affecting their introduction and diffusion are considered.

The report also covers the nature of the diffusion process, energy and development, needs of the rural and urban poor, the characteristics of the technologies, and their feasibility and acceptability by the poor. Further, developing country experience with these technologies is briefly described, followed by conclusions and recommendations. A summary of these topics follows.

DIFFUSION OF INNOVATIONS

Diffusion is the process by which innovations spread to the members of social systems. In the diffusion of technologies, both centralized and decentralized models have been identified and characterized. In practice, most diffusion efforts combine elements of both.

In this report, the term diffusion applies both to dissemination of information about a new technology and dissemination of the technology itself; for instance, new cooking stoves.

ENERGY AND DEVELOPMENT

For the rural and urban poor, properly designed biomass conversion technologies could reduce the economic and environmental costs of cooking and heating and, in some cases, provide opportunities for economic growth and employment. Biomass production technologies can slow the devastating process of deforestation and soil erosion that threatens traditional subsistence agriculture and is an obstacle to long-term economic growth.

NEEDS OF THE POOR

Meeting the energy needs of the poor through biomass-based technologies will not in itself significantly reduce a nation's petroleum use. Most of the poor already rely heavily on biomass sources--firewood, charcoal, agricultural residues, and dung--and will probably continue to do so. The value of the various technologies described here lies in increasing the availability of the materials currently in use, ensuring that they are used effectively, and providing alternative employment opportunities.

TECHNICAL FACTORS

Certain characteristics of the technologies appear to make some more acceptable than others. While many technologies are feasible, only a few may be practical. Characteristics that encourage acceptance include structural simplicity, use of familiar materials and techniques, functional discreteness, and obvious, short-term benefits.

CULTURAL AND ECONOMIC ACCEPTABILITY

In any society, existing social structures, economic organization, political institutions, and beliefs influence acceptance of change of any sort. New technologies that mesh with indigenous systems of resource allocation, work organization, goods distribution, social and authority structures, and prevailing values and religious beliefs clearly have the best chance for success.

DIFFUSION OF THE TECHNOLOGIES

Attempts have been made to introduce the technologies described to many developing countries. In almost every case, factors external to the technology seem to have a greater influence on acceptability than the technology itself.

Moreover, there are circumstances in many developing countries that prevent even initial consideration of biomass-based energy technologies; for instance, policymakers may focus entirely on large-scale hydroelectric, geothermal, or fossil fuel development schemes.

All kinds of energy sources should be considered. However, biomass production for energy has the additional potential, if properly managed, of stabilizing the environment and providing employment opportunities in rural areas.

OTHER FACTORS AFFECTING ENERGY TECHNOLOGY DIFFUSION

The diffusion of biomass energy technologies, or lack of it, is only one element of the total energy sector in developing countries--which, in turn, is only one, albeit critical, sector of national resources that must compete for attention. Developing countries generally have limited capacity to gather complete and accurate resource information, to analyze this information, and to plan allocation and use. Yet planning capacity is fundamental to rational policymaking; inability to identify clearly the nature and scope of problems, particularly where resources are limited, restricts development. Given these circumstances, it is difficult to establish priorities. Short-term, urgent problems tend to absorb a disproportionate amount of effort, although long-term problems may be even more critical.

There is another complication: many officials in developing country governments and regional organizations maintain that the attention given to renewable energy technologies by technical assistance agencies reflects an underlying policy to deprive developing countries of their fair share of oil supplies at reasonable prices and to make them dependent on "second-class" technologies. Yet even if oil prices decline further, the problems of deforestation and erosion in the countryside will continue.

Further, there are many reasons why biomass energy technologies are not considered as serious alternatives to other energy technologies. For example:

- Coal, peat, hydroelectric and geothermal sources, and oil exploration are more attractive for reasons that include abundant data on development costs, equipment, and uses, and because of offers from multinational entrepreneurs who are often on the doorstep promoting cooperative endeavors in conventional energy technology. Conversely, renewable energy technologies have no visible constituency; they are frequently associated with women's activities--firewood gathering, food processing, and cooking--and tend to receive less official attention than if they were in the male province. They are perceived as being small-scale and diffuse, and therefore ineffective or socially complex and expensive alternatives.

- There is a tendency to center on one alternative energy source rather than a multiplicity of sources, or to equate a past failure of one technique with the unsuitability of all of them.

- Solar technologies and related high technology energy systems are developed by industrial enterprises to supply energy needs in industrialized countries, and it appears unlikely that these technologies will make much of a contribution to developing countries in the foreseeable future. Their application, other than in relatively simple crop-drying systems and water heaters, is unlikely to reduce the amount of biomass required to continue to provide the main fuel source.

- There is a perceived lack of infrastructure for diffusing energy technologies. Almost everywhere, extension capabilities are felt to be inadequate, even for simple reforestation; and even though nongovernment organizations (NGOs) are frequently successful in working with traditional social agencies, few governments acknowledge that this route offers serious possibilities for easing national energy problems. Many governments fear that apolitical NGOs will not remain so.

- Biomass-based technologies are seen only as long-range solutions, since growing biomass or organizing its production on any useful scale is believed to take too long. Hence, the technologies are often given little attention either by planners or by farmers with a serious immediate problem.

- The low rate of internal return on investment that is felt to characterize these technologies pushes scarce capital or loan funds into conventional development projects in agriculture and industry that carry high short-term returns. Long-range societal gains such as environmental benefits are almost never considered in economic evaluation.

These problems appear to be almost universal, occurring at individual, community or village, and government planning levels and applying equally to arid and humid conditions. In arid regions, the energy problem and its environmental consequences are stark and visible, and to many, hopeless; in humid areas, the problem is less obvious because forest resources mask the rate of deforestation. The problem in these areas is as serious, however, because of the potential impact on watersheds, the high rate at which environmental degradation can take place, and the enormous numbers of people affected. Because environmental degradation has no direct impact on the area being deforested, there is little incentive for the woodcutters to change their practices.

Biomass Production

Biomass production for fuel has been practiced for thousands of years. It is only recently that population increases have made substantial, potentially disastrous inroads on the regenerative capacity of woody plants in agricultural areas.

Reforestation schemes have been a feature of land use for some time, but often for watershed management or timber production rather than fuel. An important feature of these projects has been the protection of the trees from use by neighboring communities. In many countries, the forestry authorities are mainly concerned with policing forest reserves, rather than with assisting the community in growing trees for fuel use. Because forestry is perceived as having little value to the community (as opposed to logging interests), most forest services are poorly staffed and equipped, separated from agricultural programs, and generally ill-prepared to meet the needs of a national biomass program. Further, the species about which these forest services are knowledgeable may not be the most suitable for fuelwood.

The exploitation of forest resources is a common source of corruption, because large profits are to be made from timber concessions. Even where there is a national policy for the conservation or generation of biomass, there is often a countervailing local interest in its exploitation.

Land tenure is a particularly thorny aspect of reforestation. Land use often implies ownership; outside efforts at reforestation are thwarted by local concern over loss of rights. In this context, fuelwood reforestation projects, which in recent years have been launched in many countries, have seldom reached their goals. Where communities become involved in planning and implementing projects and where the potential benefits are clear, fuelwood programs have worked. Recent World Bank studies have shown that under favorable conditions large-scale reforestation programs have been successful.

Improved Cooking Stoves

Much effort has been concentrated on designing improved cooking stoves, for the most part using firewood. Success has been mixed: some communities and governments have adopted the technology with enthusiasm, while others have encountered design, cultural, or cost factors that hinder widespread adoption. As yet, there is no demonstrable connection between the introduction of "improved" cooking stoves and a lowered rate of firewood consumption, though in most cases this is the justification for the effort. There is considerable variation in the way that stove efficiency is measured and reported. Indeed, in some cases the improvement in efficiency and reduction in fuel consumption attributed to the devices is intuitively assumed, not tested or measured.

While stoves clearly have other benefits for the user, there is no guarantee that they will reduce fuel consumption by the amounts predicted in laboratory efficiency tests. Much of the field information about them is derived from the relatively affluent, who can acquire the technology and for whom access to firewood is not a critical limitation. Thus, even where stoves are constructed and used as the designers intended, they may merely be used for additional purposes rather than for reducing consumption. Little information is available about the desperately poor and the actual or potential impact of improved stove technology on, for example, reducing the amount of animal dung burned. Also, scientists have demonstrated that when traditional (three-stone) cooking is done carefully, efficiencies are equal to those of well-designed chula or Lorena stoves. There is little doubt that the poor who suffer firewood shortages use their supplies with care.

Charcoal Kilns, Stoves, and Gasifiers

Charcoal kilns of many types and sizes--metal drum, retort, brick beehives, dome--have been tested in a variety of circumstances. Few have been adopted, and charcoal production in many developing countries is mostly carried on by itinerant entrepreneurs using traditional pit or mound methods. Where the wood is gathered from the commons free for the cost of labor, there is little incentive to acquire a relatively expensive device, that is difficult to transport to new production sites, to improve the efficiency of charcoal recovery.

In many countries, particularly arid ones where trees are scarce, there is a remarkably sophisticated and efficient system of contractual relationships among charcoal makers, transporters, and consumers, providing employment and income to many people at the expense of the countries' dwindling forest resources. This system has proved difficult to alter, even in extreme situations in which charcoal must be transported hundreds of kilometers. The capital investment and political impetus required to balance use with regeneration of the trees, in a way that would involve existing charcoal producers rather than displacing them, has not yet been evident in most countries. Yet, as petroleum fuels and electricity become ever more expensive, charcoal consumption in urban areas for home use is increasing dramatically. This holds true in restaurants and industry as well. The environmental consequences are potentially catastrophic; many societies are mortgaging the resources of future generations, not only placing at risk many forest species (which may become extinct along with their wildlife) but also causing erosion--all of which threaten the entire ecological base on which agricultural production depends.

Successful charcoal fuelwood plantations have been operated in a few locations, notably Brazil and Argentina; more serious attention from governments, communities, and technical assistance donors is needed for this aspect of national energy supply. Charcoal use is likely to increase even further as other technologies, such as gasifiers to generate electricity, pump irrigation water, or power fishing boats or trucks, are adopted. These technologies can provide a useful incentive for environmentally sensible planning of renewable fuelwood supplies. Charcoal conservation, through improved production techniques, briquetting, and use of more efficient stoves, is an important priority.

Biogas

Perhaps none of the renewable energy technologies has been promoted as enthusiastically as biogas generation. Yet, experience indicates that the conditions under which the technology is successful on a sustained basis are rather restricted. Biogas generation requires capital investment, a plentiful and reliable source of substrate (preferably animal dung), and a fair level of technical competence to obtain enough gas to justify the effort. Community biogas projects have foundered over problems with the cost of the system, responsibility for operation and allocation of gas, collection of substrate, disposal or equitable distribution of the residue, and, in arid regions, adequate water supply. Few poor families have the capital or the necessary animals to support a home generator. The amount of gas produced, particularly in cooler regions, is relatively small compared with the capital and running costs, even discounting labor (which is often neglected in biogas economics).

Thus, with notable exceptions, successfully operating biogas generators are typically associated with fairly sophisticated, integrated systems of waste management based on cattle, swine, or poultry production, in which the gas generated is a valuable by-product rather than the main objective. The extensive Chinese experience, with some seven million digesters constructed, appears similarly linked to environmental and health improvement and nutrient recycling (including human waste), with gas a by-product of nutrient conservation and waste treatment rather than the principal objective. In other countries, diffusion is subsidized and, in general, adoption is limited to the relatively wealthy animal owner.

It appears unlikely that biogas will provide much fuel for the poor until they acquire other resources--such as livestock--and are in fact no longer poor.

Alcohol Fuels

Only in Brazil is fuel alcohol used on any scale. However, many countries are now undertaking pilot projects to produce ethanol as a substitute for gasoline, and a few are examining wood gasification as a route to methanol. Methanol production is inherently a large-scale, expensive technology, which is unlikely to supply fuel to the rural or urban poor. Ethanol, on the other hand, is produced almost everywhere as potable spirits, and may offer an attractive and economical source of fuel for some developing countries. It requires a cheap and plentiful supply of substrate, however, and increased demand for ethanol is likely to increase the price of substrates such as sugarcane, cassava, and cereal grains.

If used wisely, ethanol can provide farm power and, where land permits, can raise incomes in rural areas through local production of fermentation substrates as cash crops. Practical biological systems to convert lignocellulose to sugars for ethanol, which may be achieved in the near future, should markedly increase the economic and environmental attractiveness of this fuel, with profound consequences for tropical countries.

CONCLUSIONS

All biomass-based energy technologies have inherent limitations in supplying national energy needs, and it is difficult for planners to make informed judgments about appropriate mixes of these technologies for different situations. However, the production of biomass is of critical importance, both for meeting energy requirements through whichever technologies are suitable and for countering massive deforestation as people increasingly meet their energy needs from unreplenished biomass. The consequences of deforestation reach far beyond availability of firewood and are already having a devastating effect on watersheds, land contour, soil fertility, and, potentially, the entire ecological basis of sustained food production.

Maintenance of the environment, revegetation, protection of forest resources, and diffusion of suitable biomass technologies are problems that are too large and complex to be tackled only by individuals and small communities. They must be the responsibility of society as a whole.

Poor people will not change their practices unless they can afford to; that is, unless they are convinced it is to their advantage, and advantage is almost always perceived in economic terms. Of course in certain cases, especially in less monetarized societies, it may be a social or cultural advantage. The overwhelming problem with the adoption and diffusion of biomass-based renewable energy technologies is that the advantages may occur so far in the future or may be so diffused throughout society--for example, as with reforestation--that they often cannot outweigh the immediate loss in the form of effort or investment (whether to plant and nourish trees, obtain the materials to construct a better stove or biogas generator, or build a kiln to make charcoal more efficiently).

The converse is also true: demonstrate the advantage, the stove, the gasifier, and the guaranteed access to land and the rights to their own trees, and the poor will grasp the opportunity without hesitation. The packaging of policies and procedures to assure that the advantages accrue in a constructive manner is a key issue in development.

There is need to allocate a much larger share of national development efforts to reforestation, social or agroforestry, and community or commercial woodlots with whatever kind of organization local conditions require. This should include providing economic incentives, such as temporary subsidies, training, and research to support the increased level of activity. There is a particular need to examine the legislative, legal, and customary aspects of land use that affect the right to use forest resources, and to take steps to ensure that these resources are protected and renewed, rather than indiscriminately exploited. In some countries, examination of land use practices has identified necessary legal changes that have already proved highly successful in encouraging biomass production.

Many aspects of biomass-based energy technologies are highly location-specific. A great deal of local experience with these technologies is required to make informed judgments about their potential to contribute to national energy budgets. Fortunately, increased attention is being given by governments and technical assistance agencies to this need for experience. However, there are still many countries, particularly the most needy, which have yet to establish such programs. Plans for the production and use of biomass-based fuels should include, with full community participation, the following elements:

- determination of the target population's needs and resources, with policies based on this data;
- identification, evaluation, and selection of strategies for biomass development, with designation of timetables;
- experimentation with, and pilot-level production of, energy through biomass technologies under actual field conditions;
- commercialization of the technologies, with environmental safeguards;
- enhancement of education and extension services;
- support for nongovernment organizations promoting renewable energy projects; and
- cooperation and communication with other countries and international organizations to capitalize on relevant experience.

Although the use of renewable energy technologies remains very limited compared with the needs, there are some striking examples of success, with clear indications of the factors responsible. These factors that promote success include:

- Awareness of national and community energy needs and of the importance, in the local situation, of developing biomass-based alternatives. Because of the long-term nature of some biomass-based systems, and the implications for policy and land-use planning and capital requirements, there must be a national policy commitment. This commitment must be reflected in the attitudes of local officials; enthusiasm on the part of nongovernment organizations and individuals or communities by itself is not sufficient.
- Community involvement in the planning, implementation, and management of local biomass energy projects, and clear understanding of the benefits and how they are to be divided. In this context, the role of the organizer-entrepreneur (whether a government official or community leader) in motivating communities to solve their energy problems has been critical.

- Access by the community to reliable and unbiased information about renewable energy technologies used elsewhere, to capital or credit, and to technical assistance for troubleshooting.

- Incorporation of benefits other than energy in the plan; for example, improved environment and health, or development of marketable skills or products. Energy is not a discrete problem for the rural poor; rather, it is one strand in the web of poverty.

- Inclusion of opportunities for adaptation, improvement, and feedback to enhance acceptance and diffusion.

These factors appear to be equally important in organizing reforestation or building biogas generators or improved cooking stoves. The organization and management of the effort appear to be more critical than the precise nature of the technology.

1 DIFFUSION OF INNOVATIONS

Diffusion is the process by which innovations spread to the members of a social system over time. Diffusion studies are concerned with messages about new ideas, whereas communication studies encompass all types of messages. Because the messages are new in the case of diffusion, there is a degree of risk and uncertainty for the recipient, leading to somewhat different behavior than if the messages were about routine ideas. The "classical model" of the diffusion of an innovation consists of the innovation, defined as an idea perceived as new by an individual, which is communicated through certain channels, over time, among members of a social system.

In this discussion of diffusion, the classical model is extended to include descriptions of several kinds of innovations.

A preventive innovation is a new idea that is adopted "now" to avoid the possible loss of a desired value in the future. Its purpose is to minimize loss. Some examples are crop insurance, seat belts, and fire alarms.

Most of the innovations studied in past diffusion research have been incremental innovations. An incremental innovation is a new idea that is adopted now to gain a possible increase in a desired value in the future; for example, a new seed variety or chemical fertilizer.

Little is known about preventive innovation decisions, except that their rate of adoption is usually slower than for incremental innovations. The motivation to adopt a preventive innovation is often a cue-to-action, that is, an event that crystallizes an attitude into overt behavior. For instance, a drainage cue-to-action would be a flood or outbreak of waterborne disease.

DIFFUSION AND SOCIAL INEQUITY

During the 1960s and 1970s, applications of the classical diffusion model to development programs in developing countries was criticized for increasing socioeconomic inequity. Critics point out that in development programs, the nature of the social system in which the innovations were introduced and the quality of the innovation have often been disregarded. They say that the socioeconomic structure has considerable effect on the innovation behavior of individuals and that it usually favors adoption of new ideas and technologies by richer people.

Most observers agree that the major problem in many developing countries today is the unequal distribution of such resources as income, lands, skills, and information that perpetuates inequality. Critics stress that technical change results in a skewed distribution of benefits; individuals who have greater resources usually benefit more from the innovations introduced by development agencies than individuals with fewer resources, thus widening the benefits gap.

In addition, some of the basic assumptions of the diffusion model have been criticized: (1) that communication by itself generates development, regardless of socioeconomic and political conditions; (2) that increased production and consumption of goods and services constitute the essence of development and that a fair distribution of income and opportunities will necessarily derive, in time; and (3) that the key to increased productivity is technological innovation, regardless of whom it may benefit or harm. Critics of these assumptions point out that the main inhibitions to development may be mainly structural rather than informative and that restructuring of a society may be needed to make the diffusion of innovations more effective in the development process.

The social structure has been found to be a powerful determinant of individuals' access and response to the mass media. The more privileged farmers who own land and enjoy a higher socioeconomic status have more communication opportunities and are the most likely to adopt new agricultural technologies. A farmer's failure to adopt innovations may be due more to a lack of opportunity than to resistance to change. Farmers with more land, money, and knowledge can more easily obtain the credit and information they need to adopt technical innovations. Most of the poorer and less progressive farmers in developing nations lack such resources.

Development agencies tend to provide assistance to a relatively small number of wealthy, educated, and information-seeking farmers; following this progressive (or "easy-to-convince") farmer strategy leads to less equitable development.

Some critics have also argued that innovations may be uncritically viewed by development workers as "good" for all farmers. The social and economic consequences for the community as a whole have not always been considered, such as whether a technological innovation is appropriate for everyone, or whether it favors some groups of individuals at the expense of others.

REINVENTION

Reinvention occurs when an innovation is changed by the adopter in the process of implementation. Reinvention ("adaptive" technology) implies a more active role for the individual adopter. Most scholars have made a distinction between invention and innovation: invention is the process by which a new idea is created or developed, while innovation is the process of adopting an existing idea.

In the classical model of diffusion the innovation was regarded as essentially unchanging in the process of its diffusion and adoption. The existence of reinvention was simply ignored, perhaps because the degree of reinvention may have been rather low in the case of important agricultural innovations like hybrid corn and 2,4-D weed spray, owing to the genetic, chemical, and biological nature of these innovations.

DECENTRALIZED VERSUS CENTRALIZED DIFFUSION

In addition to the centralized diffusion model, an alternative model of decentralized diffusion also exists. Schön pointed out that theories of diffusion have characteristically lagged behind the reality of emerging systems. He particularly criticized the classical diffusion theory, which he termed a "center-periphery model," and which rests on three basic premises:

- that the innovation to be diffused exists fully realized in its essentials, prior to diffusion;
- that diffusion is the movement of an innovation from a center out to its ultimate users; and
- that directed diffusion is a centrally managed process of dissemination, training, and provision of resources and incentives (Schön, 1971).

The traditional centralized diffusion model not only prevails in theoretical writings, but has become the "dominant normative model for diffusion." The best-known example of a center-periphery model of diffusion is the Federal Extension Service of the United States Department of Agriculture (USDA), which coordinates with the state agriculture extension services. This program (with an annual budget of more than $350 million) represents the largest public investment in a diffusion system in the world.

In their early days, extension services in the United States identified and then disseminated the agricultural practices followed by the most successful farmers. In later years, a superstructure of agricultural research was connected to agricultural extension services to serve as a source of innovations. Most extension workers tended to forget that innovations could come from farmers' experience as well as from formal research and development.

Agricultural innovations stem from R&D activities by the USDA and by state agricultural experiment stations. The agricultural extension services operate in close collaboration with these organizations through state extension specialists, who are stationed in state agricultural universities. The responsibility of these extension specialists is to convey information about agricultural innovations to county extension agents, who in turn diffuse these innovations to farmers. Thus, the agricultural extension services represent a centralized diffusion system. Not only has this diffusion model been widely copied in agricultural extension services in developing nations, but it has served as a basis for the design of several other U.S. diffusion systems, in such varied fields as vocational rehabilitation, mass transportation, energy conservation, and education.

Schön's main criticism of this centralized diffusion model is that it may apply to certain classes of innovations "lying near the periphery," but that it fails to capture the complexity of innovations that evolve as they diffuse (reinvention) and that thus originate from numerous sources (not just the center). The diffusion process should not consist only of centrally developed information, but should also be shaped by user demands and solutions to problems from other users.

In many developing countries, agricultural extension agents have low status; they are often drawn from agricultural colleges at which they have had relatively undistinguished careers, failing to be selected for scholarships for further study abroad or for the more prestigious positions in the Ministry of Agriculture. Few are from the rural areas or are particularly motivated to solve rural problems, and seldom do the training courses prepare them for communicating convincingly with farmers. They usually find that the farmers know more about agriculture than they do, and they become demoralized and ineffective. Giving them additional responsibilities for energy is unlikely to be any more successful, yet the prospect of erecting a parallel extension structure for rural energy is daunting.

An alternative model of diffusion, based on decentralized, horizontal diffusion of innovations among adopting units depends mainly on peer networks for transferring technological innovations among local-level units.

DECENTRALIZED DIFFUSION IN CHINA

Diffusion in China is characterized by the use of exemplary units (production brigades, communes, or counties) as role models for the rest of the nation and by "on-the-spot conferences" through which innovations originated by these model units are diffused to other adopting units throughout the nation. In the recent past, this innovation process was often started by broad policy directives issued "from the center"--that is from Chairman Mao himself. These directives were distilled from the ideas of the masses, rather than coming from educated experts. Mao wrote that one should "take the ideas of the masses (about their needs and possible solutions) and concentrate them (through study, into more systematic innovations); then go to the masses and explain these ideas until the masses embrace them as their own, carry them out and persist in this on their own; then test the correctness of these ideas in action." This statement, in short, reflects a decentralized type of diffusion.

The Tachai Production Brigade. The local unit that perhaps best illustrates how a nationally recognized model serves as a dissemination point in a decentralized diffusion system is the famous Tachai Production Brigade in Senshi Province, an area characterized by rocky soil, erosion, and poor agriculture. The approximately 90 households in Tachai Brigade manage 144 acres of this hilly land. By constructing stone terraces, underground conduits to carry off occasional flood waters, an irrigation system, and adopting chemical fertilizers and other agricultural innovations, the farmers of Tachai Brigade were able to raise their grain yields from 1,050 kilograms per hectare in 1949 to 5,295 in 1965-66, and 8,220 in 1977. This achievement is remarkable when compared with China's national average of 3,300 kilograms per hectare.

But the main lesson of Tachai Brigade is self-reliance. For example, after a disastrous rain in the early 1960s washed out all of the stone terraces that had been laboriously constructed by the Tachai farmers, this bridgade adopted the slogan of "three don't wants": don't want state funds for recovery, don't want state grain, don't want relief materials. This self-reliant stance was entirely consistent with Mao's philosophy, and in 1964 he proclaimed: "In agriculture, learn from Tachai." This slogan was reproduced on dams, bridges, and the walls of houses and public buildings throughout China.

Once Tachai became Mao's agricultural model, it was challenged by Mao's rivals, Liu Shao-chi and his followers, who criticized the Tachai Brigade "for putting politics in the background" and advocated their own agricultural model. After six years of struggle with the Liuists, who claimed that the Tachai farmers had falsified their accomplishments, Tachai emerged in 1970 as the national model for agricultural development. An interest in models is one strategy that Chinese leaders use to promote their favorite policies. After the model attains success, the political leader draws attention to it through the mass media and by on-the-spot conferences involving participants from other communes.

One important effect of on-the-spot conferences (in addition to providing information about the innovation being studied) is an increased feeling among the participants that they can control their own futures, rather than being controlled by fate, nature, or supernatural forces.

On-the-spot conferences allow participants to actually see the innovation in use by a local unit, to ask how effective it has been and how to implement it, and to consider how the innovation might be utilized in their home units. Following the on-the-spot conference, the participants report this information to their peers, who then decide whether or not to adopt the innovation, and, if so, how to fit it to their particular local conditions. China also used a decentralized approach in promoting biogas technology, with strong emphasis on local self-reliance.

Biogas Production in China. Biogas generation in China appears to be the world's most successful example of renewable energy technology diffusion, success being measured by numbers of people motivated and technology units constructed; it offers important lessons about the introduction and diffusion of new processes.

Although the first effort to popularize biogas use was based on the instructions of Chairman Mao during the Great Leap Forward, significant progress was not achieved until the 1970s. Two counties in Sichuan Province developed successful techniques for producing biogas between 1970 and 1972, building several hundred digesters. In 1972, leaders in the State Scientific and Technological Commission and the Ministry of Agriculture and Forestry decided to launch a national movement to popularize biogas. During the 1970s, about 7 million digesters were constructed, with at least a few in half the provinces and in over 1,000 of the nation's 2,100 counties. About 5 million digesters are located in Sichuan Province. This concentration in Sichuan is due partly to the higher population density there (up to 600 persons per square kilometer), partly to the relative lack of coal and firewood in the area, and, perhaps most important, the fact that the Party Secretary of the Province has been an enthusiastic promoter of biogas.

Given the policy emphasis on local self-reliance, the training of commune members has been a strong factor in the diffusion process for biogas. Biogas promotion officials have used reports, posters, demonstrations, seminars, and workshops for training. Moreover, teams of commune members have visited areas that are particularly advanced for on-the-spot conferences. On their return home, a few digesters are built for evaluation and, in many cases, modifications made in the design and construction to meet local needs.

The vast majority of Chinese biogas units are designed for the needs of a single household. These family-sized units are usually about ten cubic meters in volume and provide about one cubic meter of gas per day--sufficient for a family of five.

These family biogas units are enclosed domed pits with provision for adding wastes, removing digested effluent, and capturing and conveying the gas generated. The construction materials are brick, sand, cement, and rocks; material costs are about $12 per cubic meter of capacity; construction time is about 1.5 laborers per day per cubic meter. The state reimburses 20-30 percent of the materials cost to the owner of a successful digester.

The principal advantage of the Chinese digesters is their low cost and ease of construction. By contrast, the floating gas holder (usually steel) incorporated in the biogas units promoted in India significantly increases their cost and reduces their acceptability.

A problem with the Chinese diffusion of biogas units, at least partly derived from local reinvention and self-reliance, is that more than a third of those constructed are inoperable, primarily due to faulty construction.

Replicating the Chinese biogas experience in other developing countries is also problematic. Where the process is viewed solely as a way to produce energy, important elements are overlooked. Factors of health, environment, and recycling nutrients to the land are of equal or greater importance in the Chinese approach. The tradition or willingness to work with human, animal, or agricultural wastes is missing in many countries.

SELF-RELIANCE IN DECENTRALIZED DIFFUSION SYSTEMS

In the cases of China's Tachai Brigade and biogas programs, a local system responded to a nationally set agenda of priority needs. The innovative local program was then promoted by the central government through the mass media and widely emulated by other local units. The central government was an essential ingredient in these decentralized diffusion systems in (1) setting a national agenda of problems, (2) identifying a local model, and (3) promoting diffusion of the model's innovations. Therefore, such diffusion systems are not completely decentralized, although they are relatively decentralized in comparison with such centralized diffusion systems as the agricultural extension services in the United States. In China, decision making about which innovations to develop and diffuse, and to whom, is widely shared among the members of the system, rather than being concentrated in the hands of a few.

COMPARING CENTRALIZED AND DECENTRALIZED DIFFUSION SYSTEMS

Table 1.1 shows some of the main differences between centralized and decentralized diffusion systems. Most diffusion systems, however, exhibit some characteristics of both models. Further, over a period of time, a diffusion program can move on a continuum between the two extremes to become relatively more or less centralized. This flexibility has important implications for policymakers who are designing a diffusion system, enabling them to centralize or decentralize the system as they take into account such factors as the heterogeneity of the target audience and the nature of the innovations (for instance, whether the technology is relatively simple or complex).

To date, little comparative research has been conducted to determine the advantages and disadvantages of centralized and decentralized diffusion systems. Neither system is best for all situations. Instead, there may be specific conditions under which the diffusion system designed by planners should be a mixture of these two models. A first, and very important, step is to recognize that the centralized diffusion model is not the only possibility.

TABLE 1.1 Characteristics of Centralized and Decentralized Diffusion Systems

Centralized Diffusion Systems	Decentralized Diffusion Systems
Centralized control of decisions by national government officials and technical experts	Local control by community officials/leaders
Diffusion from top down from experts to local users	Peer diffusion through horizontal networks
Innovations come from formal R&D by technical experts	Innovations come from local experimentation by non-experts
Low-risk innovations with relatively high advantage	High-risk innovations with relatively low advantage
Projects that are of low priority to local organizations	Projects that are of high priority to local organizations
Technology-push, emphasizing needs created by the availability of the innovation	Technology-pull, created by locally perceived needs and site-specific problems
Low degree of local adaptation and reinvention	High degree of local adaptation and reinvention

DIFFUSION NETWORKS

One of the key elements in the decentralized diffusion systems is that of communication networks, defined as the interconnected individuals who are linked by patterned flows of information. It has long been recognized that interpersonal communication channels from peers are most important in persuading individuals to adopt new ideas.

Most individuals do not actively seek technical information from the most competent sources. Instead, they tend to seek information about technological innovations from sources that are local, easily accessible, and interpersonal (usually from neighbors).

The diffusion, or the lack of diffusion, of renewable energy technologies is influenced to a greater or lesser extent by the following groups or individuals within a country:

Policymakers, Planners, Politicians. Central government authorities play an important role in disseminating technologies. In many countries they are the key to approval of a technology initiative or to its funding. Often, however, renewable energy technologies as an entire category may not be perceived at this level as a serious alternative to conventional energy technologies. In other cases a critical decision may be taken to promote certain technologies as a matter of national priority --fuel alcohol in Brazil, for example.

Industrial Entrepreneurs. Industrial entrepreneurs may influence diffusion by adopting a technology and producing its key elements: stoves, gasifiers, and stills, for example. Their decisions affect diffusion by making essential components available, by influencing public and commercial perceptions of the technologies, and by supplying capital for their development. However, in many countries there is no incentive for entrepreneurs to become involved; there may be no profit opportunities among poor people, or no special role for entrepreneurs where many can manufacture inexpensive devices.

Provincial Government Officials. There may be a difference in perceptions of the value and feasibility of energy technologies between planners and officials at the national level and those at the provincial level, particularly concerning regional versus national priorities.

Private and Nongovernment Organizations. These voluntary organizations frequently have a key role to play because of their close contact with potential recipients and their ability to recognize needs, identify opportunities, and motivate the recipients.

Local Government Officials. These officials have an important role in diffusion because they are responsible for conveying national policy or reporting local needs to the national government. They may include important personnel such as extension agents and public health officials.

Village Leaders. Traditional social leaders may also be important in diffusion as arbiters of consensus for or against innovation. Their role in organizing community use of the technology or land resources may be crucial.

Family. In many cases perceptions at the family level will determine the extent to which a given technology will be adopted.

Individuals. There may be different responses to technologies at the individual, as opposed to the family, level. Wives may have a different perception of the need for fuelwood from husbands; wives may welcome an innovation that leads their husbands to play a larger role in helping them (running a biogas plant for the house), or may resent an intrusion into their affairs (building a mud stove).

SOCIETAL CONFLICTS

It may be difficult to identify clear-cut benefits to society from the introduction of renewable energy technologies; the social costs, even though not clearly defined, may appear to be heavy.

Neither the costs nor benefits are evenly divided among social groups; urban groups frequently are favored in their access to goods and services, and perhaps subsidized at the expense of their rural counterparts. City dwellers tend to receive more political and commercial attention than their compatriots in the rural areas who are frequently regarded as conservative and difficult to organize, motivate, or assist.

The cost of fuel in the cities, where it is almost always distributed commercially, has been rising, and city dwellers have been responding by switching to cheaper fuels hitherto found only in the country, such as firewood and charcoal. This, in turn, has driven up the price of wood and charcoal in rural areas and made them more scarce. Indirectly, greater charcoal use has led to deforestation, erosion, and the drought-flood cycles that affect increasingly large areas, with serious consequences for agriculture and rural life in general. Rural people are forced to flee the land and move to the cities, thereby increasing the urban demand for energy. Meeting immediate urban energy needs by increasing electrical capacity, importing coal or oil, or planting peri-urban fuelwood lots may offer better short-term returns, politically and economically, than equivalent investment in the rural areas.

National political goals are often at odds with local interests. National defense and industrialization priorities frequently divert government resources from rural development. Conversely, where a clear perception of the importance of self-sufficiency in fuel and decentralization of industry does exist, it may lead to local advantage at the expense of the cities.

The interest in alcohol fuels in many countries has led to recognition of the "food versus fuel" dilemma, in which scarce food grain resources may be diverted to produce fuel alcohol or good agricultural land is given over to sugarcane or other fuel feedstock. Yet rural development depends, ultimately, on providing energy for agriculture and for increasing the flow of goods and services that will stem the urban drift of rural populations (National Academy of Sciences, 1981). The challenge is to find ways to turn energy needs into opportunities for rural areas that can increase the market value of their products, their self-sufficiency, and the stability of their production system.

Development objectives may influence the context for diffusion of renewable energy technologies directly, through such policies as reducing oil imports or encouraging rural electrification in support of increased agricultural production. Other development goals may influence the climate for diffusion of these technologies in subtler ways.

REFERENCES

Beltran, L.R. 1976. Alien premises, objects and methods in Latin American communication research. Communication Research 3:15-42.

Bordenave, J.D. 1976. Communication of agricultural innovations in Latin America: the need for new models. Communication Research 3:43-62.

Chen, P. 1976. Population and Health Policy in the People's Republic of China. Occasional Monograph 9, Interdisciplinary Communication Program, Smithsonian Institution, Washington, D.C., USA.

Downs, G.W., Jr., and Mohr, L.B. 1976. Conceptual issues in the study of innovation. Administrative Science Quarterly 21:700-714.

Esman, M.J. 1974. Popular participation and feedback systems in rural development. In: Communication Strategies for Rural Development, edited by R.H. Crawford and W.B. Ward. Cornell University Press, Ithaca, New York, USA.

Mao Tse-tung. 1954. Selected Works. International Publishers, New York, New York, USA.

National Academy of Sciences. 1981. Staff Report: Energy and Agriculture. National Academy Press, Washington, D.C., USA.

Oksenberg, M. 1974. Chinese policy process and the public health issue: an arena problem. Studies in Comparative Communism 7:375-399.

Perkins, D., et al. 1977. Rural Small-Scale Industry in the People's Republic of China. University of California Press, Berkeley, California, USA.

Roeling, N.G., et al. 1976. The diffusion of innovations and the issue of equity in rural development. Communication Research 3:63-78.

Rogers, E.M. 1976. The passing of the dominant paradigm: reflection on diffusion research. In: Communication and Change: The Last Ten Years and the Next, edited by W. Schramm and D. Lerner. East-West Center, Honolulu, Hawaii, USA.

Rogers, E.M. 1962. Diffusion of Innovations. Free Press, New York, New York, USA.

Rogers, E.M., et al. 1977. Extending the Agricultural Extension Model. Institute for Communication Research, Stanford University, Stanford, California, USA.

Schön, D. 1971. Beyond the Stable State. Temple Smith, London, England.

Steidlemayer, P.K. 1975. The Da Zhai Model in Chinese Agriculture, 1964-1974. Ph.D. Thesis, Stanford University, Stanford, California, USA.

Taylor, R.P. 1981. Decentralized Renewable Energy in the People's Republic of China. Science and Technology Unit, World Bank, Washington, D.C., USA.

Warner, K. 1974. The need for some innovative concepts of innovation. Policy Sciences 5:443-451.

Zaltman, G., et al. 1973. Innovations and Organizations. John Wiley & Sons, New York, New York, USA.

2 ENERGY AND DEVELOPMENT

Effective diffusion of biomass energy technologies is vital to achieving economic growth and providing sustained food production in developing countries. The energy crisis in many developing nations is twofold, encompassing both the need to reduce oil imports for the urban and industrial sectors and to sustain--or enhance--agricultural production and the rural economy. The crisis threatens the traditional subsistence technologies and jeopardizes the fledgling industrial capacity. It is, however, a complex ecological, developmental, and international phenomenon of which oil prices and supply are only one facet. In some countries it has had abrupt and devastating consequences, while in others its effects are long-range and more subtle.

The problem of energy places the provision of food and other basic needs at risk not only because of insufficient fertilizer or mechanization, but also because of environmental degradation and deforestation caused by expanding agricultural lands and the unrelenting harvest of firewood. Possibilities of growth through economic diversification are blocked, and the natural resource base is imperiled almost everywhere.

In the traditional sector of most developing countries, the diminishing supply of firewood is the most immediate cause of this growing shortage of energy. The process has become tragically circular. Briscoe's figures on the Bangladeshi village of Ulipur are illustrative. "In the last century," he writes, "the numbers of inhabitants have increased by 350 percent, while production of food grains and fuel resources has remained essentially the same. Since cooking technology is unaltered, fuel requirements have risen with the size of the population" (Briscoe, 1979). As population growth in Ulipur, and nearly everywhere else in developing countries, outdistances increases in the productivity of traditional technologies, food can be grown in greater quantity only by shortening periods of fallow or by clearing forested land. Removal of this vegetational cover further degrades the land--by the drought-flood-erosion cycle in the drier regions and through leaching of topsoil nutrients, flooding, and siltation in more humid tropical areas. As forest and bush are cleared for farming or depleted by overgrazing, the fuelwood search extends ever further, endangering the already marginal livelihood of the rural majority.

In the urban sector the energy problem is different, but equally critical in its impact. Increasing amounts of scarce foreign exchange must be allocated to importing petroleum fuels to keep the industrial base in operation and to meet the growing domestic energy needs--for cooking, heating, lighting, and transport--of an expanding population.

For urban and rural poor alike, economic growth is blocked by the energy crisis. The potential of social unrest is increased, and for many, survival itself is threatened.

Most developing countries have few fossil fuel resources, and for most the prospects of discovering any are limited. An alternative often advocated, but not seriously heeded, is reorganization of the world system of energy resource distribution in which a greater share of petroleum fuel resources would be reallocated, or at least sold more cheaply, to needy nations, out of some so-far-undiscernible sense of global fraternity. Such a radical transformation in the international system of conventional fuels distribution (or any other resource) seems unlikely.

Stable, productive land-use systems for meeting food and fuel needs have evolved in some areas over many hundreds of years. (Indonesia)

There is, however, an alternative that warrants serious attention. It is to develop technologies for generating and using biomass-based sources of renewable energy. Reliance on these energy sources is, of course, as old as fire. What is new is the way such energy is generated and harnessed to meet human needs. Properly designed, successfully diffused--adopted, adapted, and spread--biomass-based energy technologies have the potential for reducing dependency on fossil fuels; for more efficient, economically and environmentally less costly means of cooking and heating; for providing urban dwellers with cheaper energy sources; for powering the small-scale industries essential for reducing urban unemployment; and for renewing the sources of biomass on which all these technologies depend.

Biomass energy technologies use resources that, although not evenly distributed, are available in most countries. Most are also nonpolluting. Relatively little foreign exchange is needed to acquire them; generally they can be replenished without major reliance on either external financial or material support. Melchor sees in the promotion of biomass-based energy technologies the potential for reducing income disparities that are a major source of tension in many developing countries. "The most important promise of this newfound use of biomass resource is that its very ubiquity favors social, political, and economic patterns that will encourage self-sustaining systems that carry the potential for promoting equity and the reordering of the rural-urban balance" (Melchor, 1981).

Such technologies will not only respond to developing countries' short-term energy requirements, but by fostering biomass production will also slow the devastating rate of deforestation and soil erosion that threatens long-term economic growth. Credits that now go for importing costly conventional fuels can be redirected to investing in development and restoring the ecological balance.

But few of the biomass energy technologies have been diffused on any scale among the developing country populations most in need. This is due not so much to the limitations of the technologies themselves as to a failure to adapt them to the variety of economic and sociocultural settings into which they must be introduced.

REFERENCES

Briscoe, J. 1979. The Political Economy of Energy Use in Rural Bangladesh. Environmental Systems Program, Harvard University, Cambridge, Massachusetts, USA.

Melchor, A., Jr. 1981. Options Immediately Available to the Region's Developing Countries to Lessen Their Dependence on Imported Hydrocarbon Fuels. United Nations University, Tokyo, Japan.

3 NEEDS OF THE POOR

The immediate beneficiaries of biomass energy technologies should be the poor, both rural and urban, in developing countries. Their energy needs give this study its focus, not because biomass-based technologies can entirely resolve their nations' difficulties with escalating petroleum prices, but because of the immediacy of their energy needs. Their overwhelming reliance on biomass for fuel--principally on firewood, crop residues, and dung--gives the diffusion of improved biomass-based technologies critical implications for alleviation of their energy problems.

Most rural people rely principally on human and animal muscle power in their work as subsistence farmers, herdsmen, fishermen, cash cropping smallholders, and plantation laborers. Clearing and plowing the land; planting, cultivating, and harvesting; and threshing, pounding, grinding, and storing are all done largely by hand, with some help from draft animals and a few simple tools. Most crops are transported from field to household and to the local market on foot or animals. Firewood, used mainly for cooking, is the principal fuel.

As food consumption outstrips the productive capacity of such traditional, low-energy farming systems, lands are lost through overcultivation, overgrazing, erosion, and soil exhaustion. Clearing forests for cultivation and to meet the growing urban and industrial demand for fuelwood can destabilize the natural resource base and lead to diminished agricultural productivity, intensified fuelwood shortages, and an increased need to purchase food and fuel. Consequently, a growing portion of the already reduced productive capacity of the rural poor is redirected to market sale. In many areas a single cash crop--coffee or sugarcane, sisal, tobacco, or pyrethrum--monopolizes the best remaining land and the preponderance of the labor force. When the rural population's requirements for food and fuel must be met by cash purchase, the money that farmers earn through sale of their property, produce, or labor often is inadequate. As their farmlands are degraded through overworking and erosion, or are lost altogether to meet the need for cash, their economic position progressively worsens. As herdsmen raise more livestock for market, their herds overgraze available pasturelands and they, too, grow steadily poorer.

Those who do not remain in rural areas to cultivate their overworked farmland, or who do not migrate elsewhere in the countryside in search of farm work or forested lands to clear, move in growing numbers to the towns and cities, usually to begin a frustrating search for alternative means of making a living. In the towns there generally is little industry to absorb this unskilled influx of job seekers. Urban unemployment and underemployment are rising. Even those who find work are frequently unable to earn enough for their material needs or for social services like education and health care. Morehouse and Sigurdson's (1977) pessimistic projection on employment in India typifies the situation in many developing countries. From 1961 to 1971 the number of landless agricultural workers in India, for whom these conditions are particularly acute, increased to 90 percent.

To maintain the urban poor and to contain political unrest, the governments of many developing countries, with support from assistance agencies, endeavor to hold down prices for farm commodities and fuel. This further discourages rural producers, who then join the flight to the towns, making the problems of food and fuel supply still worse.

Those whose growing energy needs might better be met through biomass energy technologies also include the small-scale entrepreneurs --food processors, artisans, manufacturers of household goods, machine parts, and export commodities--who often share some aspects of the marginality of life among the urban and rural poor. Their needs are similar, and their role is critical, for generally they have more capital. Less constrained, they are often the innovators. And it is their sector of the economy that frequently allows for that expansion and diversification of production essential to reducing unemployment and achieving economic growth. For those in the monetarized sector of the economy, the diffusion process is less complex, since it is primarily controlled by economic factors; for those in the subsistence sector, it is more complex, since it is dependent on nonmonetary social and cultural factors.

STABILITY AND IMBALANCE

Most rural people make their living by labor-intensive farming. In times of stress, gathering wild plants, fruits, leaves, nuts, berries, and roots, and hunting animals are important ancillary activities. Plants are also often a source of materials for artisans and builders or for herders' shelters; for instance, palm leaves for thatching roofs, banana fibers for making rope or twine, exudate chicle for chewing gum, copals and dammars for varnish, bark for cloth and canoes, seeds for food or oil, and roots or bark for pharmaceutical needs. In India, literally millions of people are employed for part of the year in forest gathering. When cover is cleared for new farmland or for fuelwood, the supplies are endangered.

In Bihar in central India, hundreds of arrests and many deaths have resulted from a "tree war." The controversy began in 1977 when the Bihar Forest Development Corporation started to replace the natural sal forests with teak. The peoples of Bihar use the sal for construction, eat its fruits, and sell the seed oil for industrial uses. The sal tree is also used in religious ceremonies. Teak has value only as commercial timber. In late 1980, a gathering to protest further encroachment was fired on and thirteen men were killed. An uneasy truce exists, in which teak planting has been halted for the time being.

In some areas, especially in Africa and the Middle East, subsistence cultivators live in a sometimes contentious but symbiotic relationship with neighboring herdsmen. As herds are increased in response to market incentives, the pressure on available pastureland grows. Competition with farmers over migration routes and access to pastureland increases at the same time that bonds of reciprocity, often based on the exchange of food grains for pasturage rights and dairy products, break down. As livestock trample or graze off available vegetation, farmlands continue to spread, and land is set aside for game parks, herders are forced to constrict or redirect their traditional and ecologically balanced nomadic patterns; the vegetational cover is further diminished; the fuelwood shortage is intensified; and the process of ecological devastation speeded (National Academy of Sciences, 1981).

Fishermen frequently share a parallel problem as the encroachment of farmlands diminishes the forest cover on which they rely for cooking and heating fuel, boat building materials, and food. Mangrove swamps, essential for fish breeding and shore protection, are harvested for charcoal.

The food-producing and food-gathering techniques of rural people in most developing countries and their access to markets have kept their populations at subsistence level. Food surpluses are characteristically small or may be bought by state purchasing agencies at low prices to support urban demand. The population is engaged much of the time in the quest for food. Except where intensive irrigated agriculture or commercial crop plantations are possible, opportunities to acquire wealth are limited. Generally, there has been neither the surplus wealth nor the time for such traditional societies to develop the institutional complexity often equated with modernity and progress.

Yet there is evidence that traditional patterns of subsistence, before they were thrown off balance during the colonial and postcolonial periods, were more in harmony with the natural resource base than are the modernizing societies that have succeeded them. Farmlands were left fallow longer, to regenerate naturally. In areas of permanent agriculture, trees were an integral part of home gardens or compound farms.

Herdsmen's patterns of migration were more wide ranging and less ecologically destructive. Fishermen frequently lived in a more stable relationship with neighboring farmers with whom they traded. The situation among the rural poor is one in which an equilibrium based on centuries-old, delicately calibrated patterns of land use has been badly disrupted. The firewood shortage in rural areas and the high cost of conventional fuels in urban centers is symptomatic of a population growing dangerously out of balance with its environment.

By promoting increased biomass production and greater reliance on more efficient biomass-based energy technologies at all levels, this ecological disequilibrium could be brought under control or reversed. With half the world's population dependent on biomass fuels, the need for this is urgent.

TECHNOLOGY IN CONTEXT

Successful diffusion of biomass energy technologies in developing countries depends upon understanding how the poor are accustomed to organizing their productive activities, regulating their social relationships, keeping order, and maintaining the normative beliefs that are essential to social stability everywhere. It is into this sociocultural context that new biomass energy technologies must be introduced. If

Population pressure, overgrazing, and tree harvesting for charcoal lead to devastating erosion and desertification. (Senegal River Valley, Mauritania)

they do not fit, they will not be diffused, regardless of their technological promise.

Many of the rural poor meet most of their needs, including their need for fuel, outside the money economy. Often patterns of exchange operate within the context of the kin group, through associations, or through other community structures to allocate rights to land and tools, to organize work, and to structure the distribution of the products of work.

However, money and markets are becoming more important. Cash is required to meet a growing number of subsistence needs, including fuel. For people accustomed to having most of their material wants satisfied through cooperative work with their kinsmen and neighbors, who do not customarily trade their labor for cash and who would never think of selling their lands, this new requirement for money is frequently hard to meet. For unskilled migrants from rural areas where fuelwood is still free, the need to purchase fuel for cooking and heating is often an unexpectedly harsh requirement. Nevertheless, the commercialization of energy eases the diffusion of technology to solve supply problems.

Many productive relationships among rural people in developing countries are still structured by kinship. The introduction of innovations designed to utilize biomass energy more effectively must take this into account. Rights and obligations derived from position in the family remain a major organizing principle in the life of the individual and community. Ties of kinship based on marriage or descent are often crosscut by associational links--membership in age grades, voluntary associations, and mutual aid societies--which lend added strength to the fabric of traditional society and determine the organization of most economic activities. Among the poor in urbanizing areas, such associational ties often serve as a valued alternative to familial relationships that are hard to sustain in an urban setting.

Still other factors affect the structure of the economic order. Age and gender play a major part in determining both the division of labor and the right to make economic decisions. In stratified social systems, class or caste position usually determines the worker's prerogative to control the products of his work and his place in the overall system by which economic rights are determined and benefits distributed.

The political organization generally derives its structure from the kinship-based social system that must be understood if community decision makers are to be effectively mobilized. This is particularly characteristic of Africa and Asia, though rural communities are increasingly being absorbed into national bureaucratic structures. Typically, village elders and other community leaders are selected from among the eldest responsible male members of the community's leading families. The poor in most countries play only a peripheral role in politics and are frequently excluded almost entirely from participating in those decisions--including decisions on technology and on the definition of their own economic best interest--that most affect the work they do and how it is rewarded.

The system of relationships that characterizes this sociocultural setting is invariably sanctioned by the prevailing system of beliefs and practices that defines right and wrong, states social norms, and generally justifies the existing order, usually on the basis of religion. Natural events often are perceived as the manifestation of religious or magical forces, sometimes leading to a fatalism that may lessen enthusiasm for innovations that promise to correct problems seen as beyond human control. This belief in supernaturally based systems of cause and effect can critically influence the outcome of energy interventions that fail to take these perspectives into account.

In most developing nations, these traditional societal structures are overlaid by new governmental institutions, which vary in the particulars of their form but are almost universally marked by weakness and instability. Generally these institutions lack both the sustained financial and human resources, especially the managerial and technical labor power, necessary to plan and maintain complex, capital-intensive, long-term development programs. Too much time and effort is necessarily absorbed in simply keeping up with current operations and forestalling economic and political collapse. Inefficiency and corruption lead to lack of inputs and marketing failures. Reliance on external economic support is almost invariable.

National administrations are generally dominated by males drawn largely from the elite. Politically, their powers are often democratic in principle, but authoritarian in practice. Outside major population centers, they frequently lack the personnel and material support necessary to maintain even routine administrative control. Their capacity to supervise and support development projects is generally even more limited. Because of this institutional weakness, assessment of national resources, including critical energy resources, is often very difficult. The situation is paradoxical. For, as Dickinson candidly puts it, "Developing countries, even though they are poorer than we [the developed countries] and cannot support our institutions, want to obtain for themselves the apparent result of our institutions" (Dickinson, 1977).

Sensitivity to efforts to export "appropriate" (perceived as second-rate) technologies can be keen. It is important to ensure a more accurate view of the problem. For, as Singer writes, "critics of the concept (of appropriate technology) sometimes argue that it in effect establishes two different standards and will therefore create and perpetuate a technological gap. This is a misconception. The gap exists in the fact that some countries are poor while other countries are rich. The task is to reduce or eliminate this gap--the economic gap. Different technologies will serve to reduce the economic gap and hence, ultimately, to eliminate the need for different technologies. . . . If we misdefine the problem by declaring that the gap is a technological gap and then try (disregarding the economic gap) to apply exactly the same technology to the two groups of countries, the real economic gap will widen further instead of narrowing" (Singer, 1977).

THE ISSUE OF EQUITY

The rural and urban poor in developing countries do not have equal access to their societies' available fuel resources. However, in rural areas where society is less stratified and the majority of people are engaged in subsistence farming, inequality of access to cooking fuels is less marked. Where fuelwood is abundant, most members of a rural community have more or less equal access to it. Most fuelwood is gathered in the course of other activities, going to and from the fields, to the well, to gather wild foods, or returning from market. However, where severe shortages occur, members of a family may have to travel long distances to gather wood, to the exclusion of their other activities. Each household collects wood for its own needs. Gathering fuelwood for sale is a relatively new activity, except on the outskirts of urban areas. The value of fuel is often not calculated in monetary terms. Disparities in distribution may go unperceived.

New fuelwood projects imply changes in land use and, often, controlled access. Plantations to supply urban fuel can compete with supplies from the landless poor and jeopardize their livelihood.

The degree of communal equity should not be overstated. Among many farming peoples, some herdsmen and fishermen, and especially among craft specialists, there may be considerable social stratification, even in small communities. For example, in India and Sri Lanka, and, to a lesser extent, in Indonesia, both personal position and family status in a caste-ranked system are important determinants of access to productive goods, including land and tools; the right to engage in certain economic activities; and the prerogative to share in the benefits of many economic endeavors.

In nearly all developing countries, even in those that are otherwise not highly stratified, women generally are assigned tasks by their male kinsmen or are otherwise constrained in their activities. Often they have no control over either their own labor or its product. The same is frequently true for members of socially inferior classes, castes, and ethnic groups. The result is that those who will be most affected by an energy innovation--a Guatemalan wife "given" an improved wood-burning stove by her husband, for example, or an Indian "untouchable" charged with collecting dung for a biodigester--may be excluded from the decision to undertake such innovations. This exclusion may be reflected in the way the individuals accept the changes these innovations require.

In the urban sector as well, the poor do not have equal access to economic benefits. Consequently, special measures may sometimes be necessary to ensure that the poor are given the opportunity to participate in planning and implementing biomass energy technologies intended for their benefit--technologies that can only succeed if their resistance is overcome.

In some countries whole sectors of the population, members of differing ethnic or tribal groups, may be excluded from participating equally in development programs of political regimes that are either indifferent or hostile to their interests. Regions such as the Senegal River Valley in Mauritania, West Irian, the Brazilian Northeast, and the territories of tribal or "national" minorities in India and China may fit this category. Such populations are often among the poorest in their nation, and their energy needs are frequently most severe and least likely to be met.

It is to be expected that biomass-based energy interventions, like other technological innovations, will prove to be neither egalitarian in their impact nor socially neutral. Writing on the relationship of India's modernization efforts to poverty alleviation, Jequier observes that "technology in general, and large scale modern technology in particular . . . tends to accentuate the social and economic differences between the small minority which can profit--or benefit--from it as consumers or producers and the vast majority of the population living at subsistence levels in the rural areas" (Jequier, 1976). Equity does not proceed automatically from either technological or economic growth or development. Rather, it appears to be the product of particular sets of institutional arrangements.

THE NEED FOR BIOMASS ENERGY

The need for cooking fuel is critical. Means for meeting this need include agroforestry or firewood plantations, improved wood-burning stoves, more energy-efficient techniques of food preparation, and more efficient charcoal production and use. In some circumstances, biogas and alcohol fuel production may also be used to meet rural fuel needs.

Where the potential exists for increasing food production through more extensive use of mechanized farm implements and irrigation, energy is also needed for pumping water; for food processing--milling, grinding, and winnowing; and for transport.

Herdsmen and fishermen also often face fuel shortages. In arid regions the watering of herds would be easier if mechanical rather than human energy were available to power the pumps (though overgrazing around wellheads is a potential consequence unless watering is carefully managed). Fishermen could increase their catch if a less expensive source of fuel were available for motor-driven boats.

For all those in rural areas, a less expensive source of energy is needed for domestic lighting. Energy to ensure a potable water supply is a serious limitation nearly everywhere.

The energy requirements of the urban poor are similar; fuel for cooking, bathing, and heating, for cooling and refrigeration, for lighting, and for small-scale industrial enterprises must all be acquired through cash purchase, often with meager cash reserves. For small entrepreneurs, energy is needed for the industrialization necessary to diversify the economy, increase productivity, and create employment. For governments of developing nations the spiraling cost of conventional energy is felt most acutely, because it affects industry, the essential transport and communication systems, and such government-sponsored services as health care and education.

REFERENCES

Dickinson, H. 1977. Transfer of knowledge and adoption of technologies. In: Introduction to Appropriate Technology, edited by R.J. Congdon. Rodale Press, Emmaus, Pennsylvania, USA.

Jequier, M., ed. 1976. Appropriate Technology: Problems and Promises. Organisation for Economic Cooperation and Development, Paris, France.

Morehouse, W., and Sigurdson, J. 1977. Science, technology and poverty: issues underlying the 1979 UN Conference on Science and Technology for Development. Bulletin of the Atomic Scientists 33(10):21-28.

National Academy of Sciences. 1981. Staff Report: Environmental Degradation in Mauritania. National Academy Press, Washington, D.C., USA.

Singer, H. 1977. Technologies for Basic Needs. International Labour Office, Geneva, Switzerland.

4 RENEWABLE ENERGY TECHNOLOGIES

Biogas, alcohol, and thermal gasification technologies can provide the energy to farmers for machinery used in producing and processing agricultural commodities; to local entrepreneurs for energy to power small-scale industries; and to households and local communities for cooking, heating, and lighting.

These technologies are dependent on the production and efficient use of biomass. In view of the importance of increasing the availability and ensuring the renewal of sources of biomass for energy purposes, biomass generation is treated in this report as a separate technology with distinctive diffusion characteristics.

Other renewable energy technologies have been excluded from consideration here because of the greater relevance of biomass energy technology to increasing agricultural productivity and the more immediate dependence of the poor in the developing world on biomass as an energy source. It is assumed that whatever is learned about the diffusion of biomass-based renewable energy technologies may also be applicable to dissemination of other renewable energy technologies.

BIOMASS GENERATION

Biomass, which is organic material usually of plant origin, is generated as a feedstock for transformation into usable energy, either deliberately and wholly for energy purposes or as a by-product of forestry or agricultural production. It includes the following types of material:

Fuelwood -- Wood grown wholly or mainly for energy purposes, in plantations or as a component of agroforestry or social forestry systems. Firewood is any wood used in cooking or heating fires; fuelwood is wood transformed into other types of fuel, such as methanol or charcoal.

Residues or by-products -- The material remaining after the removal of the forest or agricultural product (the timber or crop). Sawdust, logging residues, crop stalks, and milling husks are included in this category, as well as animal wastes.

Crops -- Biomass grown specifically or partially for energy purposes; sugarcane, cassava, beets, sweet potatoes, and other starchy crops that can be converted into glucose for fermentation to alcohol; soybeans, sunflowers, oil palm, etc., that can provide vegetable oil substitutes for diesel fuel.

Agroforestry, Social Forestry, and Fuelwood Lots

Agroforestry provides a diversified output of forest and agricultural products by combining agricultural crops with trees and livestock on the same unit of land. It is designed to achieve more efficient use of sunlight, moisture, and nutrients than is possible through monoculture and to maintain or improve soil quality.

Trees are often intensively cultivated in homestead areas and may be grown elsewhere on agricultural land as windbreaks, boundary markers, for erosion control, or to provide shade or green mulch. Trees may also be grown as cash crops.

Social or village forestry involves tree planting on marginal lands or areas bordering agricultural plots, roadways, homes, or canals. Fuelwood lots are usually established on larger protected plots set aside for this use by communities or governments, and employ forestry rather than agricultural techniques.

While agroforestry is an ancient practice, it is not yet a science. Research is still needed to determine which trees can be most successfully grown with which crops, which fast-growing trees are best combined with slower-growing species, which varieties can best withstand particular climatic conditions, and which can best meet the needs of local populations. Because of the multiplicity of technical and social variables entailed in agroforestry, it is, perhaps, the most complex of the biomass-based technologies under consideration. But with its immediate relevance to the issue of biomass generation, improved quality of life, and conservation, it is also the most critical and the most promising.

BIOMASS CONVERSION

Biomass is converted to useful energy through cooking stoves, charcoal kilns, gasifiers, and biogas and alcohol units.

Cooking Stoves

In many developing countries, the greatest proportion of national energy is consumed in the form of firewood or charcoal for cooking (and to a lesser extent, heating and lighting) purposes. The efficiency with which firewood and charcoal is traditionally burned is very low--perhaps 5-10 percent of the potential energy is actually transferred to the material being heated. Consequently, much effort has been directed to designing and diffusing stoves of higher efficiency, thereby reducing the amount of fuel required.

The three factors that most affect fuel use are moisture content, completeness of the combustion, and the efficiency of heat transfer. Air-dried wood yields approximately twice as much heat (on a weight basis) as freshly cut wood. Incomplete combustion, poor heat transfer, and heat wasted during heating and cooling are the principal causes of the energy inefficiency associated with open fires and the simple stoves used by the poor in most developing countries. Among some of these peoples, the smoke from open fires is recognized as an eye irritant and as the cause of respiratory ailments. Among others it is appreciated as a means of driving away insects. Among nearly all families dependent on fuelwood, time spent in the increasingly long search for fuelwood is a growing burden, especially for women and children.

Improved woodstove design requires a way to provide more efficient heat transfer to the cooking pots by enclosing the fire and regulating the air flow. Estimates of efficiency and potential energy savings are impressive, but usually unsubstantiated in the field.

De Lepeleire et al. (1981) have issued a compendium of stove designs. Over 100 models are described--many traditional and some experimental. From simple, shielded variations of three-stone fires to complex multicavity, multicontrol, monolithic units, these stoves represent a broad spectrum of candidates for improving fuel efficiency. However, the problem is to make new stoves acceptable by adapting them to meet the widely varying circumstances and cultural expectations of diverse populations. For as Ashworth and Neuendorffer have observed: "Cooking is one of the most culture-bound human activities. Food preparation, serving time and place, food flavor, and cooking participants often are established by long tradition and, therefore, are resistant to change. Social and religious customs may dictate all of the characteristics of the energy demand and may eliminate certain technology options even if energy output is . . . good. . . " (Ashworth and Neuendorffer, 1980).

Yet it is equally true that all over the world cooks have quickly and happily adopted gas, electric, or kerosene stoves, even at considerable expense. Historically, there are many examples of significant changes in diet taking place over short periods--most of the food produced in modern Africa is from nonindigenous crops or animals, and the taste for wheat-flour bread is ubiquitous. This underscores that the difficulty in diffusing "improved" stoves may be due to the lack of perceived improvement, either in convenience or efficiency.

CHARCOAL PRODUCTION

Because it has a higher heat content and burns with a hotter flame, charcoal is more energy-efficient than fuelwood (though much of the energy content of the wood is lost in the conversion process). It emits more uniform heat and relatively little smoke, weighs less, is more readily transported, and is easier to store. For these reasons it is preferred to fuelwood by most urban dwellers. Its production and sale can be a valuable source of income.

Charcoal is also capable of meeting needs that cannot be met with fuelwood, such as use as a reductant in the iron and steel industry. And certain types of charcoal products (briquettes, for example) enable the use of shavings, sawdust, and other residues from forest products. Organized charcoal production also allows the use of larger residues (such as branches) from the logging phase of forest industry.

Charcoal is produced from wood by pyrolysis (heating wood at high temperatures in the absence of oxygen). Because at best only about one-third of the pyrolyzed wood is obtained as charcoal, the technology is most efficient when the by-product gases and liquids can be collected and used.

In most developing countries, however, wood is converted by the simple covered pile or pit methods, used only to produce charcoal. The process is simple but inefficient. Organic liquids and volatile gases, potentially serviceable as a fuel for combustion engines, are lost. Metal or brick kilns are more efficient, but also more time consuming and costly to build. The still more complicated charcoal retorts needed to collect and recycle gaseous by-products most efficiently are beyond the financial reach of most small producers (Table 4.1).

Charcoal production is attractive when it can be marketed in areas of firewood shortage, where fossil fuel prices are rising, where it is technologically feasible to recover and use by-products, or where transport costs and storage of more bulky firewood make charcoal more economical than wood both as a domestic fuel and as an energy source in small-scale manufacturing.

Gasification

When a stream of air passes through a bed of red-hot burning wood or charcoal, the oxygen in the air reacts with the carbon in the fire to form carbon monoxide and carbon dioxide. If conditions are adjusted suitably, the proportion of carbon monoxide can be made sufficiently high that the gas can be used as a fuel source in internal combustion

TABLE 4.1 Characteristics of Charcoal-Making Devices

Type	Production (tons per year)	Capital Investment ($US, 1976)	Useful Life (years)	Wood Consumption (per ton of charcoal)
Earth pile	various	none	one firing	8-12
Pit kiln	various	small	1-2 years	7-8
Portable metal kiln	72	1,000	3 years	5-7
Brick kiln (Brazil type)	150	800	5 years	5.7
Continuous retort	20,000	2,000,000	30 years	3.5

Source: Hall, Barnard, and Moss, 1981.

engines or directly as a heat source. Gasification can be applied to many uses: the "gasifier" can be used to power stationary engines for irrigation or electricity generation, or can be made compact and portable to operate a car, truck, or boat engine. The technology can fill an important need in supplying fuel for motorized equipment where supplies of petroleum fuel are unavailable or too expensive.*

Biogas Generation

The production of biogas (principally methane and carbon dioxide) through anaerobic fermentation of organic wastes--crop residues and other plant material and animal and human wastes--provides a means of converting wastes into a useful fuel, while largely retaining their fertilizer value.

The least complex biogas digesters are simple enclosed pits or containers partially filled with water and organic materials. The gas produced by anaerobic fermentation of the organic matter is collected and fed through a tube or pipe to the stove. The digested sludge is periodically withdrawn for use in fields or fish ponds. Larger-scale units that allow for heating, stirring, and continuous addition of organic matter are more complex, considerably more expensive, and require a steady supply of biomass for economic operation.

---

*See National Research Council. 1983. Producer Gas: Another Fuel for Motor Transport. National Academy Press, Washington, D.C., USA.

Producer gas-powered fishing boat. (Philippines)

Producer gas-powered truck. (Philippines)

In addition to producing fuel, biogas generation treats human, animal, and other organic wastes so as to reduce their danger to health. Partially sterilized during fermentation, the residual sludge from the biogas digester contains fewer human and animal pathogens. Plant pathogens in crop residues are less likely to be transmitted to the next year's crop when the digested sludge, having lost little of the nutrient value of the original waste material, is spread on the fields as fertilizer. Through biogas technology, then, a useful fuel is produced, local sanitation is improved, and the soil nutrients are retained.

Biogas technologies have been most widely diffused in Asia, especially in India and China, but also in Korea, the Philippines, and Thailand. They appear to be most readily accepted where fuelwood is in short supply, where there is an abundance of suitable biomass and water not required for other uses, and where there is a tradition of handling and recycling manures as fertilizer. They may also be viewed primarily as a means of nutrient recycling or waste disposal, rather than primarily as sources of fuel. Principal problems with biogas generation technologies derive from their relative structural complexity and cost, consequent difficulties with financing and management, and the specific conditions of substrate C:N ratio, temperature, and pH that they require for maximum gas output.

They are most successfully employed in large-scale integrated waste-recycling systems, and their adoption at the individual household and community level in most parts of the world has been negligible.

Maya Farms in the Philippines is an example of a successful integrated agroindustrial operation. This 24-hectare complex maintains 15,000 pigs, marketing nearly twice that many annually. Every day, 7.5 tons of manure is converted to 400 cubic meters of biogas. The gas is used on the farm for powering deep-well pumps, slurry pumps, a feed mill, and refrigeration units of a packing plant. At night, surplus gas is used to generate electricity.

The liquid effluent from biogas production is used in fish ponds to provide nutrients for tilapia. Digester sludge is used as an ingredient in the pig feed, reducing feeding costs and supplying microbial protein and other nutrients to the diet.*

Alcohol Production

Alcohol fuel can be produced from a variety of biomass forms abundant in developing countries. Sugarcane, sweet sorghum, and numerous cereals and starches, including cassava, can all be used in the microbial fermentation process that produces ethanol. Ethanol is a suitable fuel for lighting and cooking, and it can be used alone to

---

*See National Research Council. 1981. Food, Fuel, and Fertilizer from Organic Wastes. National Academy Press, Washington, D.C., USA.

Gas produced from urban waste treatment is used to generate electricity. The nutrient-containing residues are returned to the fields. (Foshan, China)

power vehicles or mixed with gasoline as an octane booster. The stillage wastes that are a by-product of alcohol production can be used for animal feed or fertilizer. However, the more common practice of dumping stillage into rivers creates severe pollution problems.

The promise of alcohol fuel production as a renewable energy technology for developing countries lies in the fact that the biomass resources it requires are so widespread. But research is still needed to further reduce the cost of the relatively sophisticated processes required for converting such materials into usable sources of energy on a scale equivalent to petroleum fuels. Most alcohol fuel programs have focused neither on the role of small farmers as producers nor on meeting the energy needs of the rural poor. To date, alcohol fuel diffusion has been very limited. Indeed, only in Brazil is fuel alcohol an appreciable source of energy. However, many countries are now undertaking pilot projects to produce ethanol as a substitute for gasoline, and a few are examining wood gasification as a route to methanol. Methanol production is inherently a large-scale, expensive technology, which is unlikely to supply fuel to the rural or urban poor. Ethanol, on the other hand, is produced almost everywhere as potable spirits and may offer an attractive and economical source of fuel for some developing countries where land is available to grow the substrates, or where ethanol may be the only alternative fuel to keep equipment running in remote areas, or where petroleum fuels are prohibitively expensive.*

But this potential remains to be realized.

REFERENCES

Ashworth, J.H., and Neuendorffer, J.W. 1980. Matching Renewable Energy Systems to Village-Level Energy Needs. Solar Energy Research Institute, Golden, Colorado, USA.

De Lepeleire, G., Prasad, K.K., Verhaart, P., and Visser, P. 1981. A Woodstove Compendium. Prepared for the Technical Panel on Fuel Wood and Charcoal of the U.N. Conference on New and Renewable Sources of Energy. Wood Burning Stove Group, Eindhoven University of Technology, The Netherlands.

Hall, D.O., Barnard G.W., and Moss, P.A. 1981. Biomass for Energy in Developing Countries. Pergamon Press, New York, New York, USA.

---

*See National Research Council. 1983. Alcohol Fuels: Options for Developing Countries. National Academy Press, Washington, D.C., USA.

5 TECHNICAL FACTORS

In addition to purely technical viability, we are concerned here with acceptability by the poverty sector of most developing countries. Of interest are those energy innovations that can arrest environmental deterioration and increase productive capacity and those that may decrease the arduousness of hand labor and increase household and community amenities.

The sociocultural aspects of technology diffusion in the poverty sector are of particular importance where the adopters of the technology lie wholly, or mainly, outside the cash economy. Where the poor are included in the economy, diffusion of biomass energy technologies depends largely on the cost of fuel they can replace and the potential for profit the system offers to those diffusing the technology or using it to supply energy needs. However, most people, particularly in the rural areas of developing countries, lie outside the economy where energy and fuel is concerned. For them, adoption of biomass-based energy technologies is governed by a more complex mixture of factors--economic, to the extent that the technology places demands that conflict with disposable income or marketable goods and services, but also, preponderantly sociocultural, in the context of changing patterns of behavior and perceptions of need. The technologies can also bring the adopters into the cash economy, through producing fuel feedstocks or fuels for market, and they will then view the diffusion process differently.

These cultural considerations are discussed below and in the following chapter.

FACTORS THAT PROMOTE ACCEPTABILITY

From an analysis of the literature and the results of a number of project-sponsored site visits, the following six factors appear to enhance chances for technical and economic acceptability of biomass-based renewable energy innovations in developing countries:

1. Structural simplicity and scale

2. Use of familiar materials

3. Employment of familiar techniques

4. Functional discreteness

5. Integration with existing technology

6. Ability to meet locally perceived technical and/or economic needs within a locally acceptable time

This list has several characteristics important to its use as a framework for assessing the likely acceptance of innovation:

- The characteristics relate particularly to innovation at the level of the poor individual or community.

- The importance of the six basic factors varies according to situation or location, particularly with respect to knowledge or skill levels and in relation to current perceptions.

- Interactions among the six factors, or their mix in a given setting, may be more significant than any one or all taken separately.

- The factors are all dependent on the degree to which local control, or participation in the selection, of the technologies is organized.

Structural Simplicity and Scale

Simplicity facilitates the acquisition and maintenance of technologies and allows for readier diffusion among the poor. This quality also lessens the risk of dependence on external support for acquisition, operation, repair, and the replacement of parts, and minimizes the need for extensive capital investment. People living close to the economic margin are usually averse to any innovation that entails further risk to their already meager capital, to their often overestimated supply of available labor, or to their food supply.

Appreciation of economies of scale, a factor attractive to the large investor, is rarely shared by the rural and urban poor. For them, the struggle for daily survival frequently precludes a view of their economic best interest. Rather, their experience of enterprises involving extensive, long-term investment is more likely to have taught them that whatever benefits are ultimately derived are generally contingent upon factors beyond their control.

Similarly, the involvement of the poor in large-scale development schemes often has been characterized by exclusion from their participation in planning and implementation. As a result, they may lack understanding of the many economic factors that can affect the success or determine the failure of such ventures. What they do comprehend is their own relative powerlessness, that there are costs to them overlooked by outsiders, and that returns may not be commensurate with these costs. Such perceptions do not foster trust. Consequently, their response is not likely to be enthusiastic to biomass energy technologies that require what is perceived either as substantial long-term capital investment or a considerable reallocation of labor. Small-scale projects, however, are not generally attractive to planners and investors, for whom the costs of project administration are not proportional to the scale, and for whom larger projects are more manageable.

For reforestation projects, this suggests that planting efforts scaled either to the size of the local community or individual families are more likely to be accepted. The same holds for charcoal production. The principle as it applies to improved wood-burning stoves is clearly valid, for they are, by definition, small-scale devices, and the simplicity of their design, if low in cost, will enhance their acceptability. As for the production of alcohol fuels and biogas, their greater technological complexity and cost may render them less amenable to rural acceptance and diffusion, at least among the very poor.

For example, experience in India suggests that the complexity of even the simplest systems may impose severe cost constraints on the budgetary resources of developing countries. According to French, the 1979 installation cost of even small family-scale biogas plants in India was Rs3,000 or about $375. Providing 175 pounds of cow dung daily and an equal amount of water, plus the additional work involved in removing an average of 350 pounds of slurry from the tank each day, assuming no major mechanical breakdowns, would bring annual operating costs to about Rs100, well beyond the means of the desperately poor. To run the plant requires use of dung from a minimum of three to four cows. Since fewer than 5 percent of India's cattle owners have this many animals, ensuring command of the necessary supply of dung could be a problem for all but the wealthiest families. If we assess the cost of this technology as it relates to the poverty of Indian village communities, in economic terms, "even the structurally simplest biogas units are distinguished chiefly by the efficiency with which they digest money" (French, 1979).

Use of Familiar Materials

Employing familiar materials wherever possible enhances technological acceptability by lessening the requirement for learning. There is economic advantage in using materials with predictable properties, and they are generally cheaper and more readily accessible. If some imported materials must be used, "control of them must be available to the local community" (deWilde, 1977).

Use of local materials in constructing charcoal kilns and selecting wood for charcoal conversion will also increase acceptance and diffusion. If unfamiliar materials must be employed in the more complex manufacture of biogas digesters, this may lessen their chances of acceptability, and it is likely their introduction will require greater external financial support. On the other hand, customary materials may be used only because it is difficult or expensive to import materials such as plastic or aluminum.

Employment of Familiar Techniques

If the techniques for designing, manufacturing, and operating biomass-based energy technologies are familiar or are analogous to customary methods, the advantage is clear. Reliance on these methods minimizes the need for special training, the possible added cost of hiring outsiders, and the need for external supervision. Also, it reduces the uncomfortable sense of strangeness that is often a deterrent to acceptance of technological innovations.

For example, reforestation efforts that follow, or only slightly modify, customary agricultural planting and maintenance techniques will decrease the loss of seed and seedlings. If familiar construction techniques can be used in making newly designed stoves, or if local bricks can be used for efficient charcoal kilns, the finished product is likely to be more workable, and thus acceptable.

Functional Discreteness

If a renewable energy technology can function independently, its chances of acceptance are enhanced by avoidance of the cost or problems of adapting ancillary technologies. Also, its introduction does not entail disruption of the existing production system. However, in many situations progress may entail change rather than accommodation to the status quo.

For example, fuelwood lots probably do not meet this criterion of discrete function. While woodlots may appear to stand alone, almost invariably they entail reallocation of available land and labor. Improved systems of charcoal production similarly require a reallocation of productive goods, especially land, and are not more readily accepted. Stoves also fit this criterion better, especially when they simply replace an existing cooking and heating technology. Units to produce biogas and alcohol are structurally discrete, but their operation is complex and requires the reallocation of land and labor to produce the needed raw materials.

Improved stove used improperly--fire should be completely enclosed to control combustion rate. (Honduras)

Integration with Existing Technology

The economic advantages of an innovation that fits readily into an existing technological management system are evident. This integration minimizes the capital and/or labor costs of adapting traditional production patterns to meet the requirements of the renewable energy innovation. Conversely, where a new technology is not integrated with the existing system, the new technology, its product, and its economic consequences all require some adjustment.

For example, Ashworth and Neuendorffer cite the effect on food taste of altered forms of processing. High-speed biogas-powered food grinders produce a fine flour that tastes and bakes differently than coarse flour created by hand with a mortar and pestle. Mechanized grinding may have a profound impact on the allocation of work and free time within a community. Grinding and crushing by hand are labor-intensive and are often performed by women (Ashworth and Neuendorffer, 1980). Mechanization may lead to the transfer of this work to men or to a professional miller, without the development of any substitute productive activity for women.

A new technology that alters behavior in a way the people see as advantageous may also be more easily accepted. In Guatemala, for example, women's acceptance of the Lorena stove has been enhanced by the fact that it allows for a change in the pattern of food preparation. Shaller states that women prefer cooking, in a standing position, on an elevated surface. Several women even identified this difference from open-fire cooking as the primary advantage of the stove. The preference for elevated cooking includes a number of more special benefits: the increased comfort of standing as opposed to sitting or bending by an open fire; the greater degree of cleanliness afforded by getting pots and pans off the dirt floor and away from wandering domestic animals; and the greatly increased safety for small children (Shaller, 1979). Thus, the outsider must be wary of undue reverence for preservation of tradition for its own sake.

Fuelwood lots may or may not meet the criterion of integration with the existing technology. If they can be established on unused or marginal land, adjustment may be easier. Often, however, their development is seen as competitive with agriculture or herding. Where collective management of communal resources (grazing, for example) exists, it will facilitate organization of fuelwood lots. Expanding acreage of multipurpose crops, such as gum arabic, rubber, or mesquite, may be easier with energy needs satisfied as a second rather than primary objective. If a tradition of charcoal manufacture does not exist, a similar reallocation of productive goods may be required, particularly of labor, and some training may be necessary. Stoves can be designed to fit easily into an existing food-processing technology, so long as they allow for the preparation of staple foods and permit the use of a familiar variety of firewood. Because it is technologically more complex, the conversion of biomass to biogas and alcohol fuels involves greater capital and labor costs before functional integration with the existing technology can occur.

Meeting Locally Perceived Needs

The biomass-based renewable energy technology clearly should meet locally perceived needs. Often this is not the case; and often the problem is time. Biomass-based technologies are seen as long-range solutions, since growing biomass or organizing its production on any useful scale is believed to take too long. Hence, the technologies are often given little attention either by planners or by farmers with a serious immediate problem.

There must also be demand. The stronger this demand, the greater the chance of acceptance. No matter how apparent a technological need may be to the would-be donor, if the intended recipients do not see it clearly, the task of winning its acceptance may be impossible. People must perceive the technical efficacy of the innovation, their need for it, and their ability to afford it. Conversely, it is important that a new technology does not detract from fulfillment of an alternative function. Here a major problem is often the local population's strong interest in technological innovations that will increase their food or water supply.

If multiple needs are met, as they are by most agroforestry projects or by production of effluent for fertilizer during biogas generation, chances of acceptance are further increased.

Moreover, the economic costs entailed in acceptance of a new technology must be perceived by the poor as being clearly offset by economic benefits. And such benefits must be discerned as accruing not just to the community, the nation, or the "developing world," but to the poor themselves--and soon.

While agroforestry projects may appear to be responsive to a real need, the investment of land and labor they require does not promise short-term perceptible benefits. The payoff from improved charcoal manufacture is likely to be equally apparent and to become available sooner. Acceptance of improved stoves also depends on a people's concern to reduce labor time and/or the money required to obtain firewood. Labor and material costs must be offset both by a marked reduction in fuelwood consumption and a patent saving of labor time or money. There is little evidence that the former holds, and the noncommercial nature of most energy consumed mitigates the latter. It is likely to take still longer for the savings in money and labor time derived from alco hol or biogas production to become apparent, which may affect acceptance of these technologies where people are aware of their needs for an alternative energy source.

REFERENCES

Ashworth, J.H., and Neuendorffer, J.W. 1980. Matching Renewable Energy Systems to Village-Level Energy Needs. Solar Energy Research Institute, Golden, Colorado, USA.

deWilde, T. 1977. Some social criteria of appropriate technology. In: Introduction to Appropriate Technology, edited by R.J. Congdon. Rodale Press, Emmaus, Pennsylvania, USA.

French, D. 1979. The Economics of Renewable Energy Systems for Developing Countries. Report to the al Dir'iyyah Institute and U.S. Agency for International Development, Washington, D.C., USA.

Shaller, D.V. 1979. A socio-cultural assessment of the Lorena stove and its diffusion in highland Guatemala. In: Lorena Owner-Built Stoves, 2nd ed. Volunteers in Asia, Stanford, California, USA.

6 CULTURAL AND ECONOMIC ACCEPTABILITY

The traits of a society--its patterns of adaptation to the environment and its economic organization, social and political institutions, and beliefs system--will all affect the acceptance and diffusion of biomass-based energy technologies. The influence of these variables will have to be assessed in the context of each local culture into which a technology is to be introduced. However, some generalizations can be proposed about those aspects of society and culture that play an important role everywhere in determining a biomass energy technology's acceptance or rejection.

The following conditions appear to increase chances for acceptance:

1. Adjustment to the system for allocating productive goods

2. Compatibility with the existing work organization

3. Adaptation to existing patterns of distribution

4. Integration with the social structure

5. Accommodation to authority

6. Harmony with prevailing values and ideology

As with the six main indicators of technical feasibility, judging the effects of diffusion from any one of these cultural variables will lead to oversimplification. The interaction of these factors with each other is extremely complex.

ADJUSTMENT TO THE SYSTEM FOR ALLOCATION OF PRODUCTIVE GOODS

The acquisition and allocation of land and other forms of capital can often be critical to the acceptance of a biomass-based innovation.

The more that efforts at reforestation take into account local land ownership, gathering rights, and work input (especially by women), or the existence of communal cooperation, the greater the chances of success. Plans to promote more efficient fuelwood use through introduction of improved wood-burning stoves or improved methods of charcoal manufacture, must also take into account the system for allocating the raw materials required for such innovations. It is critical to know who holds the rights to disposal of the crop residues and other wastes from which biogas is processed or who owns the land on which sugarcane, sugar beets, and other crops are cultivated for conversion into fuel alcohol.

An understanding of the sources of capital for financing is, of course, equally critical. French argues that attention to financial cost is too often ignored because it "may not matter as long as outsiders are footing the bill. However, when renewable energy devices have to be purchased by individuals or local agencies they are more likely to be discriminating in spending their money" (French, 1979).

Although by definition the poor control few productive resources, their relationship to those who do may be pivotal to a renewable energy technology's chance of acceptance. If, for example, prevailing property relationships will block the impoverished majority from receiving project benefits, project-sponsored innovations are unlikely to be accepted by them.

Attitudes toward risk in allocating productive resources also need to be taken into account. Again an observation from French is pertinent: ". . . although we often will be unable to design a project which is 'objectively' risk-free, there are steps we can take to make a given degree of risk more acceptable. . . . Risk aversion is minimized where new techniques are closely related to familiar ones, farmers are expected to contribute labor rather than cash to the project, cooperation among farmers is encouraged, and dependence on outsiders is avoided" (French, 1979). During part of the year, however, labor can be a scarce commodity in rural areas. Farmers have none to spare for any project except their own farming.

The range of technologies that meet this criterion of low risk and high reliability at present is not very wide.

COMPATIBILITY WITH THE EXISTING WORK ORGANIZATION

An understanding of indigenous systems of work organization is important to the successful diffusion of a new technology. If the local labor force is to be effectively mobilized to accept and diffuse a biomass-based technology, it is important to know whether most of the relevant energy-related tasks are structured within the family, on the basis of associational ties, gender, age, or ethnic group, or accomplished through coercion or the payment of wages.

Often there is a contrast in work organization between the rural and the urban poor. Among many rural peoples, a significant portion of labor is scattered and is allocated on the basis of kinship ties. In urban areas the work of poor wage earners is generally more specific and individualistic and is characterized by competition rather than cooperation. However, weakened familial links are sometimes compensated for by ties to voluntary organizations, mutual aid societies, or to cooperatives or clubs, which can often serve equally well as a basis for organizing productive activities.

New methods and patterns of work organization that biomass energy interventions require can create psychological and social problems of adjustment. The pace, precision, and discipline required by the processes of biogasification, alcohol fuel production, or efficient charcoal making may be at considerable variance with the generally more diffuse work patterns of noncommercial farming, herding, or fishing. Or it may be that certain traditional patterns of work are more conducive to the acceptance of innovation. Schlegel finds, for example, that work organized along cooperative and consensual lines, with a complex division of tasks, and well-integrated into an intervillage or urban system, with economic influence widely shared, tends to be correlated with greater acceptance of renewable energy alternatives (Schlegel and Tarrant, 1980). The assumption that rural folk are usually amenable to cooperative efforts (or more amenable than urban or prosperous people) is implicit in this interpretation and taken for granted in many energy projects. But it is not necessarily true, and specific incentives or other conditions may be required to elicit cooperative production.

Where assignment of work or access to productive resources is strongly constrained by age, gender, ethnic group, or caste, these customs must be understood if local acceptance of biomass energy technologies is to be won. The relation of these customs to equality of opportunity is especially pertinent.

In stressing the importance of attention to the role of women in forestry, if their special needs are to be met, Hoskins points out that in most developing countries women are more apt to be illiterate, to be least served by extension services (especially in activities outside the home), may have the least flexibility in how they use their time, have the least mobility, and have the least financial resources. Very often planners opening new areas or reordering old ones assume that only men are "farmers" per se, and only they are given the right to take up land and receive credit or other inputs, whereas women may have been independent producers on their own land under the previous system. These facts make it imperative to examine carefully the issues that might prevent women's participation:

- specific problems women have in gaining and retaining access to land or the use of tree products;

- specific time, financial, or other constraints to be overcome to free women to participate; and

- assurances women have that they will receive benefits they value from forestry projects (Hoskins, 1979).

While men are often already organized in groups, women rarely are, and it may be more difficult to mobilize women for training or work efforts.

Reliance by a government on coercion as a means of organizing workers--through forced mobilization or politically sanctioned systems of peonage--is also a potentially significant variable. While coercion has been abandoned in many developing countries, often the memory lingers and government-sponsored or promoted work organizations are often either weak and ineffectual or unduly authoritarian. As for the role of nongovernment organizations outside the traditional mold, although they are frequently more effective, few governments acknowledge their potential in helping to solve national energy problems. Many governments fear that apolitical NGOs will not remain so.

Where most people work for wages, the would-be innovator needs to know how wages are set and what the work force regards as a fair return for labor expended.

For the promotion of tree planting, a clear understanding of who benefits from the work is critical to development of a locally acceptable plan for distributing, planting, and nurturing young trees, particularly to avoid problems in the seasonal availability of labor. For charcoal making, this may entail recognition of the traditional organization of production and of the potential such activity often provides for new forms of entrepreneurship. For promoting improved wood-burning stoves, knowing who works with the materials from which such stoves will be made and who is responsible for food preparation can be critical. The conversion of biomass into fuel entails equal sensitivity to the local division of labor, especially as the handling of waste products is frequently the prerogative of a specific group whose subordinate status may further complicate acceptance of the new technology or alternatively will involve an unpleasant form of labor, which women, children, or depressed classes will naturally assume they will be coerced to do. Even if the resource is freely available--for example, dung or commons trees--its transformation into a commercial commodity may be a difficult, complicated process.

ADAPTATION TO EXISTING PATTERNS OF DISTRIBUTION

Allowing for the way consumption goods are distributed can also be important. To take fuelwood as an illustration, an FAO report points out that most of the fuelwood used in developing countries is not bought and sold under market conditions. When it is marketed, supply and demand factors strongly affect the price. The weight and bulk of fuelwood severely limit the distance at which it can be economically marketed, with harvesting, transport, and handling constituting a major portion of the cost to the consumer. Fuelwood or charcoal are purchased mainly by people in towns and villages that have no access to forest areas, or by wealthier households; the price is often beyond the reach of the poor who are most dependent on this source of energy (Food and Agriculture Organization, 1979).

If it is contractually agreed that all who work equally in developing a community woodlot will get a comparable share of the resulting product, the poor will be more ready to accept this innovation, especially if they have traditionally been excluded from equal access to the benefits of such endeavors. For charcoal making, it will be important to know how the subordinate sector of the population assigned to such work is customarily compensated and how prices are set. Equal access, however, may mean in practice that it will be exploited to the benefit of the rich.

Availability of a viable system of transport (for example, roads, depots, and trucks) will affect chances for diffusion of most biomass energy technologies, especially in remote rural areas. For a technology such as improved wood-burning stoves, however, the problem of distribution is a different one; because stoves often are not portable, manufacture must be decentralized if they are to be widely diffused. And for biogas and alcohol fuels, the issue of acceptability among the poor often hinges on ease of access to the capital for acquiring and installing the complex units that production of such fuels usually entails.

To foster diffusion, the system of distribution must allow for the disposition of the product of the biomass energy technology at a cost that most people can meet. Because the cash investment required for importing many biomass energy innovations is far in excess of what the poor can afford, it is often imperative to utilize local resources and skills for the design and development of technologies that are more productive than the traditional ones and yet are within the reach of farmers and other poverty groups.

INTEGRATION WITH THE SOCIAL STRUCTURE

Compatibility with the existing social structure--family organization, associations, systems of stratification based on class, caste, ethnicity, or gender--is also a factor in winning acceptance and ensuring diffusion of biomass technologies.

Among most of the rural poor, kinship provides one basis for the structure of activities critical to the organization of community life, in its economic and political, as well as its social, aspects. The role of kinship in structuring the household as a productive unit is probably most important. Associational bonds, especially those provided by mutual aid and self-help groups and by secret societies, can play an important part in ordering social life at the local level. Class or caste position, ethnic identity, and gender as determinants of individual status and community organization are frequently important.

In the Sahel, for example, Ki-Zerbo recognizes the need to modify energy-inefficient cooking technologies, but points out that change must proceed with recognition of the relationship between the existing food-processing technology and women's social status. Her description of the symbolic placement of the three cooking stones typical of most Sahelian hearths illustrates the affront to women's position that an insensitive effort to introduce an alternative cooking technology could entail. An elderly woman, generally the wife's mother or aunt, usually sets up the three stones in the fireplace. Should the man displace them he shows that he repudiates his wife (Ki-Zerbo, 1980).

Noronha refers both to Burundi, where continuing bitterness between ethnic groups is likely to have a major effect on project implementation, and to India, where ethnic stratification and accompanying traditions of economic specialization also have important implications for project design. Referring to the "tribal" groups of Gujarat, which account for 14 percent of that state's population, he writes that "tribal" groups such as those who live in the main forest areas of the state not only form the main employment core of the afforestation works, but also are aware of the different uses of tree species, since they live largely by what they produce from the forest (Noronha, 1980).

ACCOMMODATION TO AUTHORITY

Understanding the local political system--the organization for making and enforcing community-wide decisions, settling disputes, and regulating relations with neighboring peoples--and how it relates to the system for installing and maintaining a biomass-based renewable energy technology will enhance the chances of adoption. For only when local decision-making processes are understood, and those with true authority are identified, can effective plans be made to ensure popular local participation and acceptance.

The relationship between the local political order and representatives of the national government must also be understood if government officials are to serve as intermediaries between external donors and the local population. The administrative task is complex. Even after the biomass energy intervention has been selected and acquired, it must be properly "packaged," combined with extension services, information and communication services, credit, and the necessary technical components. The complexity of these adminstrative tasks is compounded by the fact that they are themselves "imported"; they must be adapted to local conditions by overburdened, frequently underpaid, and often undersupported administrators whose own managerial training, obtained abroad, may be as "inappropriate" as the new energy interventions they are charged to implement (Singer, 1977).

Writing about Upper Volta, Winterbottom describes a problem of faulty administration that is a recurrent cause of project failure: extension workers' reluctance to discuss the rationale behind their program with local farmers. Rather, those responsible for local-level implementation of the government's reforestation program tried to achieve innovation without explanation: "A Neem-Cassia-Eucalyptus package . . . was presented to them as an unalterable wholly satisfactory technology . . . . An elitist attitude . . . led to the inference that communication need be only one way . . . . The general lack of local participation in decisions at the outset . . . contributed to the eventual abandonment of the project" (Winterbottom, 1979).

For agroforestry projects to succeed, the authority structure that backs up the traditional land-tenure system must be understood as well. For charcoal production, the role of political leadership in setting prices and regulating distribution can be critical. For disseminating stoves, the supportive role of government-sponsored extension workers may be essential. The political implications of biomass conversion, which often entails dependence on credit, government subsidies, and community mobilization, also must be understood. In China, close integration between the energy objectives of the national government and the local communes appears to have been critical to the diffusion of biogas generators. The remarkable spread of this technology in the People's Republic of China can be attributed to four government-supported rules:

1. the three-in-one principle in which the latrine, the pigsty, and the biogas digester are constructed as an integrated unit, enabling an ordinary family to have a biogas unit;

2. the use of local materials that suit local conditions, enabling considerable cost savings;

3. full support from the commune, the production brigade, and the production team, providing the important element of mass participation; and

4. a simple digester design and construction techniques that allowed the unit to be operated using traditional skills.

But China is exceptional. In many developing countries the infrastructure needed to implement such a system is lacking.

Even in the most noncentralized processes of diffusion, where reliance on local initiative may be of major importance, the central government still has a role to play if diffusion is to succeed. It is the government that should set an agenda of problems, identify a local

Many single-purpose digesters have been abandoned. (Guatemala)

model, and promote diffusion of the model's innovations. But even this minimal leadership role may be difficult for the governments of many developing countries to fulfill. Even if central governments were able to assume a more significant role in the promotion of biomass energy technologies, it does not follow automatically that the energy problems of those most in need would be alleviated.

HARMONY WITH PREVAILING VALUES AND IDEOLOGY

Ideology, often but not invariably in the form of religion, plays a key role in the lives of most of the world's poor. It frequently sanctions the existing economic and social systems and provides legitimacy to the political order. Local religious leaders can often use their authority effectively to foster, or block, the acceptance and diffusion of new technologies. And it is from people's beliefs that their moral values are derived, including those that may encourage or inhibit the acceptance of change. Technologies that do not pose a threat to existing ideology, to basic moral values, or to the status of religious leaders, generally have the best chance of acceptance.

REFERENCES

Food and Agriculture Organization of the United Nations. 1979. Issue Paper for the Technical Panel on Fuelwood and Charcoal. Prepared for the Preparatory Committee's Technical Panel for the United Nations Conference on New and Renewable Sources of Energy, 21-25 January 1980. FAO, Rome, Italy.

French, D. 1979. The Economics of Renewable Energy Systems for Developing Countries. Report to the al Dir'iyyah Institute and U.S. Agency for International Development, Washington, D.C., USA.

Hoskins, M. 1979. Community Participation in African Fuelwood, Production, Transformation and Utilization. Report to the Overseas Development Council and the U.S. Agency for International Development, Washington, D.C., USA.

Ki-Zerbo, J. 1980. The shortage of firewood is a daily vital concern. VITA News, Special Energy Issue 15 July.

Noronha, R. 1980. Village Woodlots: Are They a Solution? Paper prepared for the Panel on the Introduction and Diffusion of Renewable Energy Technologies. National Academy of Sciences, Washington, D.C., USA.

Schlegel, C.C., and Tarrant, J. 1980. Thinking About Energy and Rural Development: Methodological Guidelines for Socioeconomic Assessment. East-West Center, Honolulu, Hawaii, USA.

Singer, H. 1977. Technologies for Basic Needs. International Labour Office, Geneva, Switzerland.

Smith, K., and Santerre, M. 1980. Criteria for Evaluating Small-scale Rural Energy Technologies: The Flert Approach. East-West Center, Honolulu, Hawaii, USA.

Winterbottom, R. 1979. Upper Volta Koudougou Agricultural Development Project. Appraisal Report for the Forestry Sub-Program. West Africa Projects Department, World Bank, Washington, D.C., USA.

7 DIFFUSION OF THE TECHNOLOGIES

Both the literature and recent reports from the field demonstrate the relevance of the variables discussed in the preceding chapters to diffusion of biomass energy technologies in developing countries. This chapter is a discussion of these variables as they relate to each of the technologies included in the report.

FORESTRY

Structural Simplicity and Scale

Forestry efforts that are small enough to avoid extensive reliance on outside support appear to be most successful. Forestry projects should be calibrated to the ecological, economic, and cultural variations from region to region. Most developing countries lack the government infrastructure and the funding essential to centrally administer large, widely dispersed forestry projects. Where a sense of common development goals has not yet been successfully engendered at the local level, government-sponsored reforestation programs--especially those on government-controlled land--appear unlikely to gain the commitment of the local population unless extension services are greatly improved. In Gujarat, for example, a government forestry project's success relied on "small, manageable and realizable targets." And as area increased, "so did extension staff, so that the scale of operation could remain small" (Karamchandani, 1981). Similar results are reported from Korea where "an incremental or step-by-step approach was used which put emphasis on results rather than abstract ideals. Realistic village potentials were stressed in each stage of development" (Gregersen, 1982).

Use of Familiar Materials

The decision to plant locally familiar species will generally prove to be both ecologically and culturally sound. Because the local population is familiar with the growing habits of indigenous trees and the economic value of their products, less external supervision is required for introducing projects to cultivate them for biomass. There is also less risk than with newly introduced varieties, which may prove to be unexpectedly invasive, to produce smoke that is found to be odorous, toxic, or irritating, or to have other serious drawbacks. Part of the success of the Korean program is attributed to the exclusive use of species that were well known to, and valued by, the rural population (Gregersen, 1982).

On the other hand, it should not be assumed that people are unwilling to try new species. Africa has many examples of successful introduction of exotic trees (eucalyptus in Ethiopia, cocoa by the Ga and Akan in Ghana, coffee by the Chagga in Tanzania) and of vegetables, where rapid growth or some other clear benefit has been perceived.

Employment of Familiar Techniques

Because of the environmental variability, cultural heterogeneity, and the weak infrastructures characteristic of government institutions in most developing countries, familiar techniques seem clearly to be preferred in planning and implementing tree-planting projects. Local agricultural techniques should be followed wherever possible. Extension efforts should build on the local population's existing appreciation of the value of local materials as a source of food, fuel, and construction and crafts materials. The value of trees in fixing nitrogen in the soil, for shade and windbreaks, and as sources of food for animals and humans should all be stressed along with their role in maintaining ecological balance.

Functional Discreteness

Tree planting does not meet the diffusion criteria of functional discreteness. On the contrary, promotion of forest growth ties in intricately with nearly every other aspect of the local agricultural scene. The situation is paradoxical. While trees play a major role in maintaining the environmental balance essential to the success of agriculture, their planting is often viewed as competitive with food growing and traditional herding. Forestry is perceived not only as disruptive of the existing production system, but as economically less advantageous. When the short-term food needs of the rural poor are measured against the hypothetical long-term benefits of reforestation, tree planting rarely comes out ahead. This bias is usually reflected in the relative power of those government agencies responsible for agriculture and forestry. Political leaders rarely give higher priority to tree planting than to expansion of the food-producing sector. But the difficulty of integrating forestry with agriculture--and with engineering schemes as well (as when deforestation results in the silting up of dams or makes mountain roads impassable)--should be viewed as a constraint, not as an insurmountable obstacle.

Integration with Existing Technology

Local perception of the role of forests in technoenvironmental adaptation needs to be assessed and built upon. Conversely, the common perception that forest cover is at variance with increased agricultural productivity must be countered. Education through competent extension appears to be the best solution. Forestry needs to be seen not only as fostering the local economy, but as offering opportunities for its expansion and diversification through the use and sale of forest by-products (such as polewood, fruit, gums, fodder, and fiber), in addition to use of the wood for fuel. There are a number of ways in which agriculture and tree planting can be integrated. These include growing fruit trees; demarcating rice fields with trees; using bush fallow for soil regeneration; planting legume trees to maintain soil fertility and to provide fodder as well as fuelwood; and using shade trees to protect individual crops, such as coffee, and as shelterbelts and windbreaks.

Agroforestry combines tree and food crops for short- and long-term benefits. (Tanzania)

Where an agroforestry effort is, by its nature, not clearly integrated with the existing technology, supportive services become more important. For example, "if a number of farmers are expected to switch from existing crops, or even to supplement these activities, there are likely to be constraints on land and/or labor," according to deLucia and Bhatia. In such instances, "careful planning is essential to ensure adequate return to subsidize inputs, if necessary, and to provide continuous extension facilities. It will be necessary to convince farmers that sufficient long-term importance is being given to the program and that they would not be at an economic loss if they switched from their existing practices" (deLucia and Bhatia, 1980).

Meeting Local Needs within an Acceptable Period

In meeting local needs within an acceptable time, the problems with tree planting are twofold. First, tree planting is likely to be seen by the local population as less critical than their food supply problems. Second, the benefits are seen as essentially long term, despite the need for fuelwood. As Rogers notes, "reforestation campaigns often fail to reach their objectives because forest depletion is not perceived as an important local problem in the community. The depletion may have occurred very gradually, over a period of generations. The forests may not have been perceived as an important resource. Forest products may not be perceived by the public as a possible solution to recent energy shortages and/or to rising energy prices" (Rogers, 1980).

Governments and technical assistance agencies have been reluctant to allocate substantial resources to forestry or fuelwood projects because of these perceptions and a skepticism regarding their economic viability.

Yet recent World Bank studies of their investments in forestry have shown that fuelwood projects have a favorable rate of return (Spears, 1980). A selection of information from these projects is shown in Table 7.1.

Where trees are valued primarily for their shade or fruits, the issue of deferred rewards may be somewhat less critical. However, the poor majority are unlikely to be enthusiastic about a program in which resolution of a pressing food need is contingent upon the successful outcome of a poorly understood, long-term forestry program. Furthermore, as deLucia and Bhatia observe, "a villager who must walk a marginally greater distance for firewood often will not perceive the problem or at least see its significance in the same way as a national planner, who sees the aggregation of deforestation and soil erosion in national perspective" (deLucia and Bhatia, 1980).

The understandable impatience of the rural poor should be addressed through agroforestry programs that offer side benefits in the shorter term: by planting varieties with usable by-products such as fruit or fodder that can meet more immediate needs, through intercropping annual crops, or by providing interim subsidies as an incentive.

Adjustment to the System for Allocating Productive Goods

Adjusting biomass energy technologies to the prevailing system for the allocation of productive goods is a problem in nearly all developing countries. With reforestation, land tenure is a particularly thorny aspect of this adjustment. Land use often implies ownership; outside efforts to stimulate reforestation are thwarted by local concern over loss of rights. In Papua New Guinea, for example, 97 percent of the land is communally owned. Since a measure of ownership accrues to the person who cultivates the land, villagers are hesitant to allow the government to initiate reforestation programs.

TABLE 7.1 Fuelwood Projects Financed by the World Bank (1977-1980)

Country	Reforestation Area (hectares)	Approximate Cost per Hectare ($)	Economic Rate of Return (percent)
Burundi	8,500	388	17
India	52,000	884	13
Korea	120,000	250	18
Malawi	28,000	500	14
Mali	3,400	1,900	11
Niger	400	6,000*	11
Philippines	28,000	300	22
Thailand	11,000	227	13
Upper Volta	3,500	2,000	16

*Irrigated

Source: Spears, 1980.

From experience with agroforestry in India, Noronha points out that "if community woodlots are to be established, there are three conditions that should be satisfied: there must be enough agricultural land for the local community; the panchayat common should be large enough so that villagers do not perceive that the reservation of land for forestry will affect their other needs--particularly need for grazing; and, third, the area set aside should be large enough for the harvest to meet the needs of all villagers . . ." (Noronha, 1980). Similar concerns have been reported in Honduras (Forman, 1981).

Difficulties with access to land, labor, and water are major constraints to tree planting in many regions. Where the economic benefits of forestry efforts are perceived as being largely deferred until the trees are cut for fuelwood, local populations are often interested only if they feel certain they will share in the projects' eventual benefits.

While much rural land in developing countries is held collectively --usually by the kin group that works it--often all rights to land, especially to land that is not under cultivation, are held residually by the national government. This can sharply reduce local farmers' commitment to participation in forestry projects. In Tanzania, for example, "abolition of freehold land in 1963 and the vesting of land in the state gives the farmer no ownership and consequently [he] perceives his holding as temporary and is unconvinced that he will reap the benefits of tree crops eight to ten years after planting" (Noronha, 1980). And legally such lands often can be allocated for reforestation without the consent of the local population. The policing of forestry projects is generally undertaken by representatives of the national government. This may create resentment and fear of government-sponsored reforestation efforts for those rural dwellers who are landless.

Government allocation of parcels of forest land often appears to work best, but here, especially, the rights of those who work such forest plots must be made explicit. In Gujarat, for example, family groups or landless peasants are allocated parcels of forest land, are paid a monthly remuneration for carrying out all operations from nursery planting to felling, and are guaranteed "a 20 percent share in the net profits when the plantation is harvested after a period of 15 years" (Karamchandani, 1981). As part of a rural electrification program in the Philippines, 100-hectare plots are leased to farmers' associations for a 25-year term (renewable for another 25 years). While growing firewood to be sold to power the generating station, the farmers are encouraged to intercrop with annual vegetables and fruit trees for a shorter-term return.

A further problem in allocating land for agroforestry is related to quasi-public areas such as roadsides, canal banks, and railwaysides. Both the right to allocate such land and the authority to administer agroforestry efforts upon it are often ill-defined. As a result, these areas are generally underutilized.

Compatibility with the Existing Work Organization

Given the weak infrastructure of most government institutions at the rural level and their tendency to be authoritarian in their approach to the rural poor, there is a need to build on local forms of work organization in planning, implementing, and maintaining forestry projects. Those who have been responsible for the care of trees and for collecting their products are those who need to be mobilized if agroforestry efforts are to succeed. Since women are the ones most often responsible for food preparation and fuelwood collection, they have an interest that could be used to persuade them to undertake the required tasks. Yet their perception of their needs and how these might best be met are frequently ignored. Projects designed to introduce new sources of firewood may disrupt or destroy existing systems, to the disadvantage of those who depend on them.

Another constraint on the organization of work for successful forestry is noted by Noronha: "The Indian village," he writes, "unlike the Korean, is heterogeneous. There is a mosaic of castes which come together only on well-recognized traditional occasions, such as weddings or temple festivals. But even on these occasions each caste plays a traditionally defined role. There is no tradition of growing trees that falls within the rubric of traditional common action. It is a tradition that has yet to be built up, and in some areas may prove difficult, if not impossible, to develop" (Noronha, 1980).

Adaptation to Existing Patterns of Distribution

When adapting the dispersal of forestry products to existing patterns of distribution, again the critical factor appears to be ensuring that those who participate in forestry efforts will be among the beneficiaries. In Gujarat, for example, profits from roadside forest plantings are split equally between the government and adjacent villages through the local community council, or panchayat (Karamchandani, 1981). In Korea, sales from plantation production are shared among cooperative members in proportion to the time each has spent on the project.

There must also, of course, be an appropriate system for transporting that portion of the wood designated for the market. But there is danger that building roads to facilitate such transport may further destroy the existing forest cover. Ease of transport is also a factor. Crooked or thorny tree species are obviously less desirable in areas where firewood must be transported long distances. The fact that firewood is generally regarded as a free good must also be taken into account. Producers who anticipate (perhaps inevitably) that firewood prices will be manipulated to favor the urban population are likely to be wary of investing their energies in forestry efforts. Put another way, with the imbalance in income levels between the urban and rural population, wood cut cheaply in the countryside may be consumed in the cities, leaving rural dwellers worse off than before.

Repeatedly, proponents of forestry projects encounter the problem of local distrust described for Tanzania. "Few villagers believed that wood would be distributed free at harvest. They felt that government or some rich villagers would get all the benefits; since all the inputs were provided free of cost by government, the villagers did not identify with the project and the free provision of inputs reinforced their belief that it was a government project in which they would not share" (Noronha, 1980).

Integration with the Social Structure

The success of forestry projects is likely to be affected by the indigenous social structure where the subordinate status of some sectors of the population--women, descendants of slaves, members of particular minority groups--are by tradition denied equal status in community economic endeavors. For example, referring to a village survey on tree planting in Rajasthan, Agarwal notes that "questions were addressed exclusively to the male heads of households whereas it is women and children who are usually the wood gatherers. If women had been asked, a different listing of priorities for planting trees might have emerged" (Agarwal, 1980). Karamchandani reports that in village panchayats in Gujarat responsible for forestry projects, "provision is made at all levels for participation of women, scheduled castes, and tribes" (Karamchandani, 1981). In Korea, women's groups have often taken the responsibility for raising planting stock in nurseries. The sale of these seedlings provides them with an additional source of income.

Accommodation to the System of Authority

Fuelwood programs almost invariably entail some reallocation of land. This can cause disputes, especially when local concepts of land tenure are not clearly and equitably incorporated with new national laws. To minimize conflict, those responsible for adjudicating land disputes need to be included, brought into the projects at the planning stage so they can work with, rather than in opposition to, those implementing government forestry policy.

Noronha ascribes Chinese success in forestry ("over a period of thirty years, from 1949 to 1979, the forest area of China increased by 72 million hectares, representing 12.7 percent of the land area") to the "effective political mass mobilization of rural communities. National targets were translated into specific goals for districts, brigades, and communes at the lowest levels; people were expected to display the same revolutionary fervor in tree planting as in, for instance, building communal dikes; the spirit pervading the program, as in all other programs, was the re-building of China under the leadership of Mao. The organization for this purpose was already in place; the division of tasks prescribed uniformly throughout the nation and implemented by a hierarchy of officials and cadres" (Noronha, 1980).

Congruence between the local authority structure and the organization of government at the national level is easiest to achieve in culturally homogeneous societies. For example, the Korean forestry development program is focused on improvement of life at the local community level within the context of the "new community movement," saemaul undong, a nationwide effort at rural revitalization that began in the 1970s and that derives its philosophical sanctions from Confucianism and Buddhism. Forestry projects are executed by village forestry cooperatives. From 1973 to 1978 more than one million hectares of land were planted by about 20,000 of these cooperatives (Gregersen, 1982).

Unfortunately, a relationship between the local population and higher systems of authority, particularly at the national level, more often than not is lacking. In many countries, forestry authorities are more concerned with policing forest reserves than with assisting the community in growing trees for fuel use. As a consequence of this conservation responsibility, in which forestry is perceived as having little value to the community at large (as opposed to logging interests), most forest services are ill-prepared to meet the needs of a national biomass program. Even species with which these forest services have experience and about which they are knowledgeable may not be the most suitable for fuelwood. The exploitation of forest resources is a very common source of corruption because large profits are to be made from timber concessions. Even in cases where there is a national will for the conservation or generation of biomass, there is often a countervailing state or local interest in resource utilization and income generation.

Another political factor that must be taken into account--perhaps despite the formal structure of the local authority system--is the role of dynamic individual leadership. Noronha offers two examples of Indian panchayats involved in tree planting; in the first, there was dynamic leadership. "A panchayat chairman who was trusted, started by planting trees on his own land and then convinced the villagers that they should have their own woodlot. Some of his answers were enlightening. 'Why,' I asked, 'did you decide on a self-help scheme rather than one where the Department would do everthing for you?' The reply: 'Because I refuse to pay the Department 50 percent of the profits for work on our land which is, in any event, an overpayment.' And then, where did he get funds, since his panchayat was poor? 'People trust me; when I need money, I approach rich villagers who give me anything I ask for.' In another self-help village, the process and answers were similar. The temple caretaker established a plantation on temple lands to obtain funds to restore his Sixteenth Century temple. The villagers saw that it was a profitable venture, and the Youth Association approached the panchayat to donate lands for a plantation which was established with the labor of all members of the Association. The Association hopes to raise money from the plantation to supplement funds to build a village school" (Noronha, 1980).

The success of nongovernment organizations with strong entrepreneurial leadership in tackling family planning and village development in Thailand (Asian Center for Population and Community Development) and Sri Lanka (Sarvodaya Shramadana), which are now incorporating rural energy in their programs, attests to the effectiveness of this approach.

Harmony with Prevailing Values and Ideology

Where religious leaders also hold political power, or where authority is powerfully sanctioned by ideology, as in China, the role of leaders in the allocation of land and the adjudication of land disputes must be taken into account. Indeed, in the Chinese case the roles of politics and ideology are inextricably linked. Noronha observes that the success of the Chinese experiments can be understood through the existence of a widespread philosophy backed by an organizational framework: that is, social philosophy realized through a political revolution, with unity symbolized in an all-knowing leader, and an organizational framework--with no little penal force--that reached down to the humblest village. Further, while observing that the Chinese case may not be an "example that could be easily imitated, it usefully points to the need both for a consistent organization and the presence of authority which, though not necessarily blatant or overt, lies behind every decision taken" (Noronha, 1980).

While he accepts the value of such ideological impetus, Noronha is careful to underscore that it does not by itself assure success. Other critical factors are cultural homogeneity, egalitarian social structure (at least at the village level), and perhaps most important, government support in the form of free seedlings, technical expertise down to the village level, continued education, and the enforcement of laws.

IMPROVED COOKING STOVES

The successful diffusion of improved cooking stoves appears to be dependent on the same roster of variables.

Structural Simplicity and Scale

Obviously, those stoves that are simplest to construct, in addition to being fuel-efficient, suitable for indoor family cooking, and easy to operate, are most likely to be copied and thus most readily diffused. If their basic design is simple, they can be more readily adapted to meet local needs and preferences. They are also easier to acquire and to maintain.

The Lorena stove, developed in highland Guatemala in the mid-1970s and widely diffused in both rural and urban Honduras, has this virtue. A simple, boxlike stove, it is constructed of clay, sand, and water.

Another simple model, a small bucket-shaped charcoal stove made in Papua New Guinea, is being successfully diffused through the distribution of a mold that can be used with clay-sand-cement mixtures to make the stoves on site. Such stoves are the least costly and can in most instances be readily moved or reconstructed. However, at the outset, a considerable financial outlay, much of it for extension, appears to be necessary for even the least complicated stoves.

Paradoxically, small-scale development projects may create administrative difficulties for funding agencies, for whom the costs of administering projects may be comparable for projects costing $10,000 or $10 million. In some countries, for instance Senegal, diffusion of improved stoves has been "packaged," so that the project may involve several hundred thousand stoves. In contrast to the successful adoption of stove diffusion by the Senegal government, other countries have perceived such an effort as too minor to warrant serious government attention, and private and voluntary organizations have in many instances taken the lead. In Ethiopia, the diffusion process has been accelerated by training extension workers to use a stove mold with which they can

An experimental clay-cement charcoal stove. (Papua New Guinea)

turn out many stoves on the spot, rather than constructing each one individually or teaching villagers to build their own.

Use of Familiar Materials

The use of local materials keeps down costs, minimizes the need for external assistance and construction, and facilitates repair. Even the use of such simple materials as cement and galvanized iron sheeting can significantly slow diffusion (Elmendorf, 1980). But this condition will vary. In Fiji, it is clay that is hard to come by for stove construction; cement is cheap and readily available (Weir and Richolson, 1980). When durable materials and shelter are not available, stoves may require frequent rebuilding.

Employment of Familiar Techniques

If familiar techniques in design, manufacture, maintenance, and operation are employed, the behavioral adjustments necessary to successful stove adaptation will be minimized and extension work with stove builders and users may be reduced; the risks of faulty construction are lessened; and repair is made easier. By allowing for adaptation to familiar technology, it is easier to accommodate the varying needs of communities. In constructing simpler stoves, rural peoples are often more familiar with what is required than are urban dwellers, accustomed to working for pay at specific tasks. For example, Elmendorf notes that the diffusion of the Lorena stove in Tegucigalpa is difficult because people in the city do not work the land, nor do they usually build their own houses of adobe or wattle-and-daub, and so they have much less of a feel for working with mud mixes. He concludes: "Where adoption of new stove manufacturing techniques is required, perhaps the most important factor in obtaining acceptance is . . . local participation in analysis, design, evaluation and implementation activities . . . . The target population should be involved, initially, as a source of information on fuel use, traditional stove design and cooking habits . . . this interaction should be used to determine design objectives for any new stove, and [to] evaluate the efficiency, usefulness and ease of operation . . ." (Elmendorf, 1980).

Functional Discreteness

Of the several biomass energy technologies under discussion, stoves probably meet the criterion of functional discreteness best. However, certain related changes must usually also be made, including modified cooking techniques and the use of different implements, pots, and other containers, and new methods of gathering, cutting, drying, and storing firewood may have to be developed.

Integration with Existing Technology

Integration with traditional methods of food preparation and fuel collection and storage will enhance chances of acceptance. Where such integration is lacking, there is a risk that the stove will be modified by users for better technological "fit" in ways that diminish its efficiency. For example, Elmendorf cites instances in Honduras where the firebox damper was removed to facilitate the burning of long pieces of wood, a change that, by introducing more air into the combustion chamber, greatly accelerated the rate of heat loss through the chimney. Weir and Richolson encountered the same problem in Fiji.

However, efforts to adapt the stoves will--and should--occur. They underscore the importance of monitoring of local adaptations by extension workers well past the implementation stage. The result may often be a new and superior mix of traditional and modern cooking techniques. In Upper Volta, for example, women who use their new charcoal stoves for smaller batches of quickly cooked food still turn to the three-stone fire to prepare large pots of slow-cooking foods (Hoskins, 1981).

In designing stoves, this principle of integration with the existing technology must be tempered by recognition that some changes in behavior may be perceived by the local population as desirable. A stove that allows cooking in a more comfortable position, that emits less smoke, or that shortens cooking time may be seen as advantageous even though it alters some time-honored methods of food preparation.

Meeting Locally Perceived Needs

For improved stoves to be diffused, the local population must think that their cooking methods are generally wasteful of fuel and labor. But firewood collection and food preparation are usually women's work. In countries where men make the decisions about money outlay, other forms of technological innovation often may be given higher priority than those designed to lessen inconvenience or discomfort to the cook. Where there is an actual demand for improved wood-burning stoves, hesitation in adopting them is more likely to be for economic than technical reasons. Because stoves are not seen as generating income (except to entrepreneurs producing them), cost of construction is likely to be the greatest obstacle to diffusion.

It should be recognized that locally perceived needs may not be reconciled with national objectives or technical assistance goals. While stove diffusion has frequently been justified in terms of saving firewood (and has been successfully carried out in the initial stages as a result of government, community, and technical assistance enthusiasm), it has become clear that fuel saving is seldom sufficient motivation for continued use of the stoves. Indeed, there is evidence that fuel saving does not occur because the stove is not properly used, not properly designed, or that it is used for purposes that were not possible with the traditional method (such as keeping hot water available), with the result that there is no net saving of firewood even though the new stove has greater efficiency (Hoskins, 1981).

Among the really poor rural villagers, improved efficiency must compete with conflicting needs. In Northern Ghana stoves were initially accepted because the villagers did not wish to disappoint the Peace Corps volunteer. After the rainy season, few were in use, since the stove had to be built outdoors because there was no room in the house and locally available materials did not permit weatherproofing; if shelter materials had been available, they would have been used to build additional sleeping accommodations, not a kitchen. Further, the stove was not perceived as particularly convenient, since it required more attention to operate correctly than the three-stone fire (Caster, 1981). In the Sahel and elsewhere, more studies of firewood use in traditional and improved stoves are being carried out, and preliminary results indicate that traditional techniques can be as efficient as the "improved" technology (Malawi Energy Unit, 1981; Prasad, 1980; Siwatibau, 1981).

Adjustment to the System for Allocating Productive Goods

In many rural areas, firewood is still a free good. If the amount of labor for firewood collection is not inordinate, adjusting to new ways of cooking may not be seen as worth the trouble.

Compatibility with the Existing Work Organization

Where firewood collection and food preparation are women's work and stove construction is categorized as a male prerogative, motivation to accept improved stoves may be lacking. Efforts at diffusion in such instances may be less successful than if stove construction were women's work, as it may be where women are potters, or where other relevant craft specializations are not closed to them.

Adaptation to Existing Patterns of Distribution

The more localized the manufacture and distribution of stoves, the more widespread their diffusion is likely to be, since they can be better adapted to meet local conditions. Conversely, dependence on a complex system of centralized production will raise distribution costs and decrease the adaptability of the stoves to local needs. So far, it appears that stoves diffuse more readily among those who are already comparatively well-off. A distribution mechanism for getting them into the kitchens of the destitute remains to be developed. An ancillary problem for this group is that more efficient mud stoves simply will not fit in their kitchens. Much of the diffusion information is directed at the relatively affluent, those who have sufficient resources to acquire a convenient, and perhaps attractively fashionable, amenity and for whom access to fuelwood is not a critical limitation.

Integration with the Social Structure

The relationship of cookstove diffusion to the social structure is exemplified by the problem resulting from women's traditionally subordinate social status. For example, it is "a major failing of researchers not to recognize the cook as an essential part of the cast of research experts" (Hoskins, 1981). Male decision makers give women's needs lower priority. This can affect both the decision to accept such stoves and the efficiency with which they are adapted to women's cooking habits and preferences. It may be important to know the way women perceive and value the chores of gathering fuelwood and preparing food in keeping with culturally sanctioned patterns. These attitudes cannot be guessed at by outsiders; they should be learned from the potential stove users themselves, by identifying and tapping those groups through which women are able to express their views--age-groups, voluntary associations including mothers' clubs, women's cooperatives, women's political branches, secret societies, sodalities, and health clinics.

Accommodation to the System of Authority

Where local society is stratified, an effective means of diffusing stoves among the poor might be by introducing them to the rich. Where such stoves are seen as more "modern," as more closely resembling the cooking style of admired urban dwellers, their acceptance may be enhanced.

Where government agencies, either national or local, are involved in the dissemination of cookstoves, the importance of accommodating to prevailing systems of authority can be seen with particular clarity. If the decision to accept new stoves is made by men, and if such new stoves are then adapted by men, with female users largely excluded, the success of their diffusion is likely to be lessened.

Harmony with Prevailing Values and Ideology

Accord with local values and ideology is particularly relevant to the diffusion of improved stoves where religion requires, for example, that women be secluded, and where the kitchen is literally out of bounds to all males who are not members of the family. In such cases, the task of those seeking to win women's acceptance of the new cooking implements may be greater. Or if the organization of the family hearth has religious significance, as it often does, this must be taken into account. However, the symbolic significance of traditional cooking methods as an impediment to diffusion should not be overstated. Referring to the three-stone stove as the symbol of a united family, Hoskins (1981) points out that "women are willing to experiment with new stoves. If a stove really saves as much as half the daily fuelwood supply, and does not require undesirable tradeoffs, what woman," she asks, "would choose the extra hour of hauling wood?" The convenience of electric, gas, or kerosene stoves has led to their widespread adoption even in relatively poor kitchens, in spite of considerable capital and recurring fuel costs.

It is important to emphasize that none of these variables represents an insoluble obstacle to cookstove diffusion. But only by taking them into account can impediments to dissemination be lessened and the chances for successful diffusion increased.

CHARCOAL PRODUCTION

Structural Simplicity and Scale

With charcoal production, the simplest technique, the earthen pit and mound kilns covered with leaves and damp soil, has diffused relatively easily where there are adequate economic incentives. The only external support required is a tie-in with a system of distribution. Frequently, as in Mauritania, charcoal production and distribution is organized by transport entrepreneurs (National Academy of Sciences, 1981). But the advantages of technical simplicity must be balanced against very low efficiency; charcoal yields are low in these simple kilns, and most of the by-products, combustible gases, and volatile oils are lost.

Traditional charcoal making. (Mauritania)

The question of scale, however, is complicated by the problem of fuelwood supply. One of the most successful projects produces charcoal on a large scale for the steel industry in Brazil. One hundred and thirty-five thousand hectares of eucalypts are harvested in five-year rotations and converted into charcoal in brick beehive kilns. The project is organized in decentralized fashion, with individuals responsible for 100-hectare forest units, including planting trees and building the kilns. In many other countries introduction of this technology, on any scale, has been frustrated by the problem of organizing a constant supply of fuelwood.

Use of Familiar Materials

Greater structural complexity, the use of brick kilns, for example, increases producers' costs considerably. Whatever its deficiencies, the earthen pit requires only materials that are available wherever charcoal production is practicable--a supply of wood and a soil bed. Where brick, cement, or metal kilns are employed, the cost rises and the possibility of unsubsidized diffusion among the poorest sector decreases.

Employment of Familiar Techniques

Even where charcoal has not been previously produced, the techniques for the manufacture of charcoal kilns are simple enough to

Tree plantations and charcoal production. (Brazil)

facilitate diffusion wherever a market exists for a cheaper, cleaner fuel. To increase productive efficiency, improving traditional methods appears preferable to imposing more complex, externally derived techniques of charcoal manufacture. In Papua New Guinea, for example, substitution of relatively inexpensive, locally available drums for earthen pits increased productivity from 20 kilograms to 50-100 kilograms per man-day and has greatly increased user interest. Current efforts are aimed at increasing the life of this system by protecting or reinforcing the drums. Brazilian experience has demonstrated the preference for smaller brick beehive kilns over larger-scale more sophisticated technology, because the wood can be cut and loaded by hand and transported to the kilns by donkeys (Florestal Acesita, 1982).

Functional Discreteness

Charcoal production meets the criteria of functional discreteness, especially when the technology is kept simple.

Integration with the Existing Technology

The substitution of charcoal for wood as a fuel source entails little modification of the traditional technology, except that a simple stove may be required. Only with the construction of more complex kilns for charcoal production is a significant innovation in the technology required, with more extensive training in construction, operation, and maintenance. Thus, greater efficiency must be balanced against higher capital costs and the need for teaching new techniques.

Charcoal manufacturing is often well integrated with traditional technology, but paradoxically, in ways that seriously jeopardize biomass regeneration processes. In many countries, particularly arid ones where trees are scarce, there is a remarkably sophisticated and efficient system of contractual relationships among charcoal makers, transporters, and consumers, providing employment and income to many people but at the expense of the countries' dwindling forest resources. Balancing fuelwood use with regeneration, in a way that would involve existing charcoal producers rather than displacing them, has not yet been seriously attempted by most countries.

Adjustment to the System for Allocating Productive Goods

No significant land area is necessary for operating charcoal kilns. If suitable trees are available, local people rarely perceive a problem in turning forest cover into fuel for charcoal kilns. There is, of course, an ecological cost, as uncontrolled tree cutting degrades the entire ecosystem. But for those likely to accept or reject charcoal production as a technological innovation, this is unlikely to be a decisive factor.

Governments, communities, and technical assistance donors need to give more attention to this aspect of national energy supply. As urban populations grow and efforts at small-scale industrialization increase, the problem of deforestation threatens to become worse. For charcoal use is likely to increase even further as other technologies--such as producer gas generators to generate electricity or power fishing boats, trucks, or irrigation pumps--are adopted.

This will undoubtedly require reallocation of land use and labor to convert the present exploitive system into one that can sustain itself. Powering trucks, boats, pumps, and electric generators with producer gas where petroleum fuel is unaffordable is an interesting possibility, one that has already been acted upon in some countries, notably the Philippines.

Compatibility with the Traditional Work Organization

Unless a wholly new system of fuelwood supply and conversion technology is contemplated, no major reorganization of traditional patterns of work is ordinarily entailed for charcoal production, except that labor may be drawn away from food production. While this is not likely to be perceived by the rural population as an obstacle to diffusion, it may increase the dependency of rural families on cash income to purchase food. In the periurban areas in which charcoal is often produced, it frequently provides one of the few forms of wage labor available to unskilled urban migrants.

Adaptation to Existing Patterns of Distribution

Among all the biomass energy technologies discussed in this report, adaptation to existing patterns of distribution is likely to be most critical to the diffusion of charcoal production. There must be a means of getting the charcoal to market, and at a price that serves as an adequate incentive to its producers. For the rural majority, the cost of charcoal is still the factor most likely to block diffusion, especially where fuelwood is still a free good.

The conditions required for profitable distribution are illustrated in a recent analysis of factors motivating charcoal production. For rural households to benefit directly from charcoal production, they must be able to buy it, and at favorable prices in comparison to other fuels, especially firewood. However, this possibility is unlikely in most rural communities in developing countries. Or, rural producers must be able to sell charcoal to the urban sector at prices that compensate the social costs incurred in producing it. Both of these possibilities should be carefully considered in developing a community fuelwood program with a charcoal component (Wood et al., 1979).

Obviously, charcoal production is most likely to diffuse rapidly where there is an established market. As the population of urban areas grows, the demand for charcoal as an alternative to firewood will increase. And as there are more and more rural merchants, governments, and teachers who do not make their living from agriculture, the demand for charcoal is likely to increase in the countryside as well. In many urban areas the increasing cost of traditional urban fuels, like electricity, bottled gas, or LPG, is causing families and businesses (such as bakeries, brick manufacturers, hotels, and restaurants) to switch to charcoal. This is further encouraged by government policies of conservation on imported fuel oil.

Where middlemen control the distribution of charcoal, there may often be a serious question as to whether producers receive a just price (Hayes, 1981).

Integration with the Social Structure

Assurance of an adequate return to those who produce charcoal is especially important in traditional societies where its manufacture is assigned to a particular, often subordinate, group. Here the issue of equity must be addressed if an increase in charcoal production is truly to benefit those whose energy needs are greatest, including many who may produce charcoal but cannot afford to purchase it. And as charcoal production diffuses among small-scale entrepreneurs with a ready supply of cheap labor, it will be difficult to ensure that development of this new industry does not exacerbate existing socioeconomic inequities.

Accommodation to the System of Authority

Since it is often the government that sets or regulates prices, charcoal prices must be established and maintained at levels that give producers the incentive to increase production.

Harmony with Prevailing Values and Ideology

Diffusion of charcoal production appears to relate to religious beliefs and practices only in those societies that assign the task of charcoal production on the basis of caste. Where others are excluded, this could affect the availability of labor necessary for the more widespread diffusion of charcoal production.

BIOGAS GENERATION

As an inexpensive source of energy from readily available organic materials, biogas generation is especially promising for the humid tropics. But the sociocultural constraints to the diffusion of biogas technology are considerable. So far, the diffusion of biodigesters has been most successful in China, a nation characterized by marked cultural homogeneity and an effective mix of political centralization and local administrative efficiency. Unfortunately, both attributes are strikingly absent in most poor developing countries.

Structural Simplicity and Scale

Even the simplest biodigester requires greater structural complexity than most other biomass energy technologies. In China the brick, sand, cement, and rock necessary to construct water-pressure family digesters at one commune is reported to cost $12 per cubic meter of capacity; 20-30 percent of this cost was reimbursable by the government. Even where financing is available, risks resulting from problems of construction persist. This is illustrated by the fate of five million biodigesters disseminated in Sichuan province in the mid-1970s, many of which are now out of operation because of construction defects. The processes of effluent treatment, storage, and transport require both an investment beyond the capacity of many would-be users and a level of technical skill that would require extensive training in most developing countries, especially where there is no cultural tradition of processing organic wastes.

Use of Familiar Materials

While local materials can often be used for most elements entailed in the construction of biogas units, the requisite supplies of biomass or water may not be available among the poor.

Employment of Familiar Techniques

In much of the developing world, techniques for the design, manufacture, and operation of biogas generators are not familiar outside a few experimental centers. Only in parts of Asia is there an indigenous tradition of biogas production that would facilitate diffusion. In most areas not only are the requisite manufacturing techniques unfamiliar but, to the extent they entail use of organic waste, especially excreta, they are potentially repugnant (Prakasam, 1981).

Functional Discreteness

Biogas production lacks the functional discreteness that appears to foster diffusion. The complexity of one system in China illustrates this: an urban power station uses domestic sewage dumped from trucks and handcarts into a gravity flow system that carries it to twenty-eight digesters. Simultaneously, an equal volume of supernatant liquid overflows to a storage tank. This material is drained into fertilizer boats that are used to transfer the effluent to nearby farms. The infrastructure to maintain such a complex and interdependent technology appears to be present mostly in nations with an established "hydraulic" cultural tradition, often dating back centuries.

Integration with Existing Technology

Adaptation would also be easier among farmers and agricultural peoples already producing crop residues and other organic wastes suitable for biogas production. In those parts of Asia where organic waste collection is an established tradition--in China and Korea, for example --this factor appears to have greatly enhanced chances for diffusion. Successfully operating biogas generators are typically associated with fairly sophisticated integrated systems of waste management based on cattle, swine, or poultry production such as the Maya Farm operation in the Philippines, in which the gas generated is a valuable by-product rather than the main product. The extensive Chinese experience, with some seven million digesters constructed, appears to be similarly linked to environmental and health improvement and nutrient recycling (including human waste), with gas being a co-product rather than the principal objective.

Meeting Locally Perceived Needs

Biogas production also fails to meet a locally perceived technological or economic need in most developing countries. Even in China, although farm managers were aware of the potential for biogas production in their farming systems, they were more concerned with increasing soil fertility and improving sanitation and health than with biogas generation. However, elsewhere in China where fuelwood is scarce, there are many communes in which 80-90 percent of cooking is done with biogas.

While biogas has a role to play as a source of fuel for small-scale industry, its potential is not familiar to the poor majority in most urban or rural areas. Familiarizing them with its benefits will entail both considerable capital input and extension support, elements already in short supply in most developing countries. Cost of the units is also a major constraint. In Fiji, for example, it is estimated that comparatively simple digesters proposed for construction at various piggeries will cost about $600 each. This investment will require a subsidy of about $300 per digester, since a capital outlay of this magnitude is far beyond the reach of the potential users. The diffusion of biogas units in this price range will probably be economically feasible only where government financing is available.

Because of its indirect benefits, Prakasam maintains that analysis of the economics of biogas generation is sometimes difficult. He says that in India, if too low a cash value is placed on the indirect benefits of cooking with biogas, such as the leisure that a housewife may enjoy or a decrease in the number of eye ailments, the net benefit to the society may be grossly underestimated (Prakasam, 1981).

Adjustment to the System for Allocating Productive Goods

Biogas production may require some reorganization of the system for allocating capital and other productive goods. In many instances labor used in food production may have to be reallocated to support biogas production, and this may prove difficult to justify.

Compatibility with the Traditional Work Organization

An aspect of traditional work organization that is perhaps most critical to the diffusion of biogas production technology relates to the handling of organic wastes, especially excreta. In Papua New Guinea and in India, such work is seen as contaminating, but in China, no stigma is attached. There a work organization based on the household unit and modified by the imposition of communal organizations appears to have lent itself well to the construction and maintenance of biogas digesters.

Adaptation to Existing Patterns of Distribution

Adaptation to local patterns of distribution is especially critical for community biogas production. Biogas projects have foundered over the cost of the system, responsibility for operation and allocation of gas, collection of substrate, equitable distribution of the residue, and, in arid regions, adequate water supply. And few poor families have the capital or the number of animals necessary to support a home generator.

In India, only households with a minimum of three to five head of cattle can aspire to install a biogas plant. At least this many cattle are needed to provide enough dung to produce biogas for the cooking and lighting purposes of an average family of four to five persons. "It is estimated that about 26 percent of the rural population in India is landless and sustained on wages earned by working inside the village or in a town close by. Certainly this population cannot afford a biogas plant" (Prakasam, 1981).

Forman's report on biogas diffusion in Colombia makes a similar point on the often prohibitive costs of a system (Forman, 1981).

Integration with the Social Structure

The Indian biogas program was restructured on income-distribution grounds because benefits were going to the relatively wealthy who had the cattle, land, and credit needed to build and use a biogas plant, while introduction of the plants by increasing the value of dung had a negative impact on the real income of the poorer groups who cook with dung. For most developing countries the question of the benefits of biogas generation to the poor remains, at best, moot.

A market-based evaluation of the benefits of conversion to biogas generation requires an often complex process of calculation. ". . . Evaluation of a community biogas plant in India shows that the benefit from biogas used for cooking depends on whether cow dung or crop residues, thus released, have alternative uses in the system. If crop residues cannot be used as fodder or sold for fuel, the benefit from biogas used in cooking will be zero. Similarly, if cow dung released from its use as dung cakes could not be marketed and the farmers did not perceive any benefits from additional organic manure or rain-fed crops, the benefits from cow dung would also be zero. In this context, reallocation of biogas from cooking to irrigation pumping or flour milling would constitute an increase in social benefits and may also make the unit financially more viable. Alternatively, the benefits from biogas used for cooking could be estimated as savings in the resource costs of providing alternative fuels for cooking . . ." (deLucia and Bhatia, 1980).

Accommodation to the System of Authority

Successful biogas generation requires the mobilization of capital and the reorganization of a significant sector of the labor force. It is therefore essential that there be, both at national and local levels, an authority system adequate to the task of administering the changes in patterns of production, distribution, and fuel consumption that are entailed by the adoption of biogas technology. Extension services that reach rural communities and a tradition of using organic wastes are two factors critical to biogas diffusion. The first of these is often absent. Prakasam, for example, ascribes to infrastructural weakness many of the problems of biogas diffusion in India. He observes: "There is lack of coordination among these various organizations. Delays are caused in obtaining loans from the nationalized banks because of too much bureaucracy. A systematic and well managed marketing system is seriously lacking and needs to be developed urgently. Follow-up work by biogas workers is lacking or inadequate in many villages, when once the biogas plant is installed. This causes serious problems to the owners of biogas plants and creates mistrust among the villagers . . . where repairs need to be made on biogas plants, services are lacking and this causes frustration to owners and creates a negative image for biogas plants and thus a resistance to accept them by others . . ." (Prakasam, 1981).

Harmony with Prevailing Values and Ideology

Attitudes toward fermentation processes in Islamic regions, and concepts of ritual contamination in many cultures, are prominent among the ideological factors that may hinder biogas diffusion. For example, "drying food with gas derived from dung may not be acceptable to certain cultures" (Ashworth and Neuendorffer, 1980). Concern about contamination is well known in India, where the handling of waste materials is relegated to members of particular subordinate castes. Perhaps partly as a consequence of this attitude, the construction of biogas plants has not yet had a "discernible impact on energy resources" there (Prakasam, 1981). DeLucia and Bhatia concur that "attitudes toward using human excreta to produce biogas for cooking, or toward carrying slurry from a plant where human excreta has been used as an input, may cause problems for villages where untouchability and casteism predominate" (deLucia and Bhatia, 1980). The concept exists elsewhere; for example, most Papua New Guinea natives are reported to have strong fears concerning the use of their excreta by others (Newcombe, 1981), and similar objections exist in Colombia (Forman, 1981).

ALCOHOL PRODUCTION

Of all the biomass energy technologies so far considered, fuel alcohol production appears to be most vulnerable to sociocultural and political constraints on its diffusion. Several countries have adopted policies to use alcohols to replace imported petroleum fuels. Only in

Large-scale ethanol plant produces 1,300,000 liters per day from sugarcane. (Brazil)

Brazil, however, has this been effected to any degree, probably because in Brazil there exists the (perhaps unique) combination of an established use of the technology, available land and labor resources to produce the feedstock, and clear government policy and subsidy.

Structural Simplicity and Scale

Alcohol production technology generally fails the test of structural simplicity. While potable alcohol production can be small-scale and the technology is relatively simple and well known, economies of scale are persuasive for ethanol and, particularly, methanol. The amount of fuel obtained from small stills requires a disproportionally large amount of equipment, effort, and perhaps most important, energy to produce it, compared with large-scale systems. Even more difficult may be the problem of distinguishing between fuel and potable ethanol produced on a small scale, and most countries have long-standing legal and tax impediments to the use of this technology.

Use of Familiar Materials

While appropriate biomass is available in many developing countries--sugarcane, sweet sorghum, cereals, starches such as sago and cassava--its conversion to fuel ethanol in quantity is rarely possible using familiar techniques, unless there is a tradition of producing potable spirits commercially from these sources. Where this technology exists, its successful expansion for fuel production may require the diversion of food staples or the land on which they are grown, with important consequences for food availability and price.

Functional Discreteness

As a technology, alcohol production is not functionally discrete, since it is dependent on the large-scale and continuous production of biomass, supplies of fuel and water, disposal of stillage residues, and distribution of the product.

Integration with Existing Technology

Alcohol production is only integrated with the local technology to the extent that forms of biomass suitable for fermentation are often available crops, although cultivated for different ends.

Meeting Locally Perceived Needs

Except in those instances where national planning allows for both high risk and the deferment of economic rewards, the high initial costs of alcohol production are likely to offset its potential long-term advantages as a source of fuel savings. Only where such disincentives are offset by substantial subsidies is this technology likely to be diffused.

Adjustment to the System for Allocating Productive Goods

Production of fuel alcohol is rarely adjusted to the system for allocating productive resources. Rather, land and labor used for food crop production must often be reallocated, with some risk that the food supplies of the producers may be decreased. Unless carefully controlled, alcohols may be used by the wealthy to fuel automobiles at the expense of food production and other land use. In Colombia and Brazil, there is concern about displacement of agricultural land for alcohol crops.

Compatibility with the Existing Work Organization

Whatever patterns of work organization are already in place appear to be adaptable to production of the biomass from which alcohol is produced. The process of fermentation and distillation, however, may entail some changes, as well as training and extensive supervision.

Adaptation to Existing Patterns of Distribution

In distributing alcohol fuels and the profits it brings, two factors must be taken into account. First, the rural and urban poor must be able to profit from its production, which means ensuring that an adequate price is paid for the biomass they produce and for their labor as alcohol producers. Second, in stratified societies there is a danger that reliance on low-cost, labor-intensive systems for producing this new source of fuel may widen the income gap between unskilled rural producers of the "raw" biomass upon which alcohol production depends and the entrepreneurs who make the investment necessary to profit from its processing and distribution.

Accommodation to the System of Authority

Because the technology is complex, a centralized authority system appears essential to coordinating the reallocation of capital, land, and labor that alcohol production requires.

Harmony with Prevailing Values and Ideology

Ideologically, there will be constraints on alcohol fuel production in those developing countries where a strict Moslem regime may oppose reliance on fermentation or where temperance interests may complicate the licensing and supervision of fuel ethanol production. Diversion of ethanol for potable use can be countered by adding small amounts of substances that make it undrinkable, such as gasoline.

REFERENCES

Agarwal, B. 1980. The Woodfuel Problem and the Diffusion of Rural Innovations. Final Report submitted to the Tropical Products Institute, United Kingdom. Science Policy Research Unit, University of Sussex, Brighton, United Kingdom.

Ashworth, J.H., and Neuendorffer, J.W. 1980. Matching Renewable Energy Systems to Village-Level Energy Needs. Solar Energy Research Institute, Golden, Colorado, USA.

Caster, S. 1981. Eight months after the introduction of improved smokeless mud stoves in the Gilibaga Sissala area of Tumu District. Peace Corps, Tumu, Ghana.

deLucia, R.J., and Bhatia, R. 1980. Economics of Renewable Energy Technologies in Rural Third World: A Review. Prepared for the Panel on the Introduction and Diffusion of Renewable Energy Technologies. National Academy of Sciences, Washington, D.C., USA.

Elmendorf, C.L. 1980. Case Study on Fechovil's Program of Improved Stove Diffusion in Honduras. Foundation for Cooperative Housing, Washington, D.C., USA.

Florestal Acesita, S.A. 1982. State of the Art Report on Charcoal Production in Brazil. Submitted to U.S. Department of Agriculture, Forest Service, funded by U.S. Agency for International Development, Washington, D.C., USA.

Forman, S. 1981. Trip Report to Colombia and Honduras. Prepared for the Panel on the Introduction and Diffusion of Renewable Energy Technologies. National Academy of Sciences, Washington, D.C., USA.

French, D. 1979. The Economics of Renewable Energy Systems for Developing Countries. Report to the al Dir'iyyah Institute and U.S. Agency for International Development, Washington, D.C., USA.

Gregerson, H.M. 1982. Village Forestry Development in the Republic of Korea: A Case Study. Food and Agriculture Organization of the United Nations, Rome, Italy.

Hayes, P. 1981. Rural Energy Technological Change and Competing End-Uses. Prepared for the Panel on the Introduction and Diffusion of Renewable Energy Technologies. National Academy of Sciences, Washington, D.C., USA.

Hoskins, M. 1981. Communication and energy: community participation in forestry projects. Development Communication Report No. 33.

Karamchandani, K.P. 1981. Gujarat Social Forestry: A Case Study. Food and Agriculture Organization of the United Nations, Rome, Italy.

Malawi Energy Unit. 1981. Energy Notes 1(2). Ministry of Agriculture, Lilongwe, Malawi.

Martin, L.H. 1980. The Ecology and Economics of Cooking Fuels in Ghana. M.A. Thesis, American University, Washington, D.C., USA.

National Academy of Sciences. 1981. Staff Report: Environmental Degradation in Mauritania. National Academy Press, Washington, D.C., USA.

Newcombe, K. 1981. Technology assessment and policy: examples from Papua New Guinea. International Social Science Journal 33(3):495-507.

Noronha, R. 1980. Village Woodlots: Are they a Solution? Paper prepared for the Panel on the Introduction and Diffusion of Renewable Energy Technologies. National Academy of Sciences, Washington, D.C., USA.

Prakasam, T.B.S. 1981. Some aspects of Biogas Technology in India and Egypt. Prepared for the Panel on the Introduction and Diffusion of Renewable Energy Technologies. National Academy of Sciences, Washington, D.C., USA.

Prasad, K.K. 1981. Some Studies on Open Fires, Shielded Fires and Heavy Stoves. Woodburning Stove Group, Eindhoven University of Technology and Division of Technology for Society, Apeldoorn, The Netherlands.

Prasad, K.K. 1980. Some Performances Tests on Open Fires and the Family Cooker. Woodburning Stove Group, Eindhoven University of Technology and Division of Technology for Society, Apeldoorn, The Netherlands.

Rogers, E.M. 1980. The Diffusion of Technological Innovations: Application to Renewable Energy Resources in Developing Nations. Prepared for the Panel on the Introduction and Diffusion of Renewable Energy Technologies. National Academy of Sciences, Washington, D.C., USA.

Siwatibau, S. 1981. Rural Energy in Fiji: A Survey of Domestic Rural Energy Use and Potential. International Development Research Centre, Ottawa, Canada.

Spears, J.S. 1980. Overcoming Constraints to Increased Investment in Forestry. Prepared for the Eleventh Commonwealth Forestry Conference, September 1980, Trinidad. World Bank, Washington, D.C., USA.

Weir, A., and Richolson, J. 1980. The Nasinu Wood-Burning Stoves: A Preliminary Report. Fiji Forestry Department and the University of the South Pacific, Suva, Fiji.

Wood, D., et al. 1979. The Socio-Economic and Environmental Context of Fuelwood Use in Rural Communities of Developing Countries: Issues and Guidelines for Community Fuelwood Packages. U.S. Agency for International Development, Washington, D.C., USA.

8 CONCLUSIONS AND RECOMMENDATIONS

When the prices of petroleum fuels rise above the cost of biomass-based energy, the production of energy in tropical forests presents an opportunity to those whose energy costs are directly linked to petroleum fuel prices. These conditions have already occurred in many developing countries. However, among the poor, who traditionally have not paid for commercial energy and indeed are often outside the money economy altogether, many other factors already discussed affect their willingness to adopt new or more efficient biomass-based energy technologies.

In this analysis of the diffusion of these technologies among the rural and urban poor, the following principal points emerged:

- Increased production of biomass is essential both to meet the energy needs of developing countries and to counter deforestation that threatens agricultural productivity and ecological balance.

- Among the variables identified as important to diffusion of a technology, allocation of productive resources (primarily land and labor), integration with traditional technology, accommodation to equitable systems of distribution, local perception of energy needs, and congruence with systems of authority appear to be the most critical.

- Organization of diffusion strategies is as important as the efficacy of the technologies themselves. The relationships among biomass resource depletion, agriculture, and energy problems are not generally understood and receive insufficient attention.

- Agroforestry and improved charcoal production and use are the most promising energy technologies of those studied.

- Biogas generation requires integrated resource management, which is usually found only in well-organized communities or large-scale systems.

- The potential competition for arable land associated with large-scale alcohol fuel production may prevent its adoption in many areas. However, this could change dramatically when practical systems to convert lignocellulose in wood to alcohol become available. Small-scale fuel alcohol production is not yet practical.

DIFFUSION OF INFORMATION

Although the focus of this report is on the diffusion of renewable energy technologies for the benefit of the poor, analysis of the factors affecting this process indicates that decisions are made at all levels of society (and indeed outside of the society, by international development technical assistance agencies and multinational corporations) that affect diffusion in many ways.

At the highest level, policymakers and planners may require evidence that any type of biomass-based technology is practical and economically feasible. Successful diffusion almost always involves a national commitment.

Development-assistance and funding agencies cannot independently institute or support energy technology diffusion projects where there is little or no national receptivity. Where a national commitment exists, the agencies also must be convinced of the economic justification of the project and its relative importance.

Private and voluntary organizations and government extension agencies have related roles to play in the diffusion process, since both are in direct contact with the intended recipients. However, they may have different strengths; private organizations usually are more attuned to local perceptions and sensitivities, but may have limited technical knowledge and resources, while government agencies may have adequate information and resources but limited sensitivity.

A principal conclusion, therefore, concerns the manner in which information about the technologies is compiled and conveyed to planners and policymakers, government and private agencies, and funding organizations.

Policymakers and Planners

At the policy level, the information must relate mainly to national macroeconomic issues. It should convey the potential of renewable energy technologies for reducing oil imports, increasing employment, and improving agricultural production--that is, benefits at the national level. For example:

- Solutions to the rural energy problem must be developed in the context of a national strategy because of the direct impact on agriculture of uncontrolled deforestation, erosion, and ecological imbalance.

- Investment in reforestation for fuelwood and charcoal, when properly managed, can generate an attractive internal rate of return and will not require a perpetual subsidy at the expense of other sectors.

- Benefits and limitations of the various technologies should be identified.

- Opportunities for private sector commercialization of these technologies should be examined.

Policymakers need less information on the technical aspects, though their advisory staff, from whom they will request technical policy options, will need more detailed information.

Development Assistance and Funding Agencies

Assistance agencies require predominantly economic information on the returns from investment in the proposed projects. To make assessments of funding needs for these technologies, however, technical feasibility studies will be required to provide data on benefits and returns at both the national and community level. Technical and sociocultural details, in addition to economic data, will be necessary. For example:

- Valid economic data should be obtained from projects that have been successful in similar situations.

- Technical entrepreneurs and private and voluntary organizations should be involved in working with local communities.

- Successful local adaptations, developed through field projects, are important to counter risk aversion in the target population.

- Emphasis on local participation must go beyond pious acknowledgment of its importance.

- Continuing technical guidance and warranties and service for commercial equipment must be available.

For these agencies, the information needs to be provided in a less general, more practical form, with allowance, wherever possible, for updating as additional case studies, economic information, or new technological developments become available. Limitations in the applications of technologies also need to be clearly spelled out. Computerized information, data books, and fact sheets may be especially useful to this audience.

Extension Agencies

Operation guidelines for government agencies and private and voluntary organizations should stress the following points:

- A survey of locally perceived energy needs should be among the first steps in planning.
- Representatives of the local population, the donor agency, and the implementing group should collaborate in planning and implementation.
- Planning and implementation should allow for regular monitoring and the early identification of problems.
- The probable sociocultural consequences of biomass-based energy technologies should be studied.
- Demonstration in a neutral context should precede wide dissemination.
- The social status of initial innovators may be critical to successful diffusion.
- Subsidies or short-term incentives should be provided when the benefits of a technology are deferred.
- In many cases, energy technologies may have to be labor-intensive and require minimal local capital investment. Tasks should be structured in accordance with local patterns of work organization, capitalizing on existing technical skills and ingenuity.
- Existing management structures--rather than imposed organizational models--should be employed, so long as equity and efficiency are not jeopardized.
- Women should be explicitly assured of equal participation in the benefits of these projects whenever possible.
- The need to minimize costs may justify acceptance of lower levels of efficiency.

- Existing systems for the allocation of land should be analyzed for possible impediments to technology diffusion.

Information on renewable energy projects must flow from extension agencies to their funding sponsors, and also among these agencies themselves. Technical journals are excellent for this purpose, but slow; newsletters are particularly helpful. The use of audiovisual aids, particularly videotapes, may be justified to circumvent language barriers, as well as to communicate excitement and enthusiasm along with the technical information.

BIBLIOGRAPHY

GENERAL

Abelson, P.H. 1982. Energy and chemicals from trees. Science 215(4538):1.

Abelson, P.H. 1980. Energy from biomass. Science 208(4450):1.

Abrahams, A.E. 1980. Renewable sources of energy and the tropical connection: an agronomic approach. In: Proceedings of the First Inter-American Conference on Renewable Sources of Energy. New Orleans, Louisiana, USA.

Adelman, I., and Morris, C.T. 1962. Society, Politics and Economic Development: A Quantitative Approach. Johns Hopkins University Press, Baltimore, Maryland, USA.

Agarwal, B. 1980. The Woodfuel Problem and the Diffusion of Rural Innovations. Final report submitted to the Tropical Products Institute, United Kingdom. Science Policy Research Unit, University of Sussex, Brighton, United Kingdom.

Ahmed, W. 1974. Constraints and Requirements to Increase Women's Participation in Integrated Rural Development. Paper presented at the Seminar on the Role of Women in Integrated Rural Development with Emphasis on Population Problems. Food and Agriculture Organization of the United Nations, Rome, Italy.

American Association for the Advancement of Science. 1979. Women and Development. Final report of a workshop conducted by the AAAS for the U.S. Department of State, 26-27 March 1979, Washington, D.C., USA.

Arnold, J.E.M. 1979. World energy and rural communities. Natural Resources Forum 3(3):229-252.

Ashworth, J.H. 1979. Renewable Energy Sources for the World's Poor: A Review of Current International Development Assistance Programs. Solar Energy Research Institute, Golden, Colorado, USA.

Ashworth, J.H., and Neuendorffer, J.W. 1980. Matching Renewable Energy Systems to Village-Level Energy Needs. Solar Energy Research Institute, Golden, Colorado, USA.

Atje, R., Bajracharya, D., Donovan, D., Koppel, B., and Tarrant, J. 1980. Energy Analysis in Rural Regions: Studies in Indonesia, Nepal, and the Philippines. East-West Center, Honolulu, Hawaii, USA.

Bach, W., and Mathaus, W.H. 1979. Exploring alternative energy strategies. Energy 4:711-722.

Bajracharya, D. 1980. Fuelwood and Food Needs versus Deforestation: An Energy Study of a Hill Village Panchayat in Eastern Nepal. East-West Resources Systems Institute, Honolulu, Hawaii, USA.

Baranson, J. 1967. Technology for Underdeveloped Areas. Pergamon Press, New York, New York, USA.

Bassan, E., ed. 1981. Global Energy in Transition, Environmental Aspects of New and Renewable Sources for Development. Sierra Club International Earthcare Center, New York, New York, USA.

Barnett, H.G. 1953. Innovation, the Basis of Cultural Change. McGraw Hill, New York, New York, USA.

Bene, J.C., Beale, J.C., and Marshall, H.B. 1978. Energy from Biomass for Developing Countries. International Development Research Centre, Ottawa, Canada.

Bernard, H.R., and Pelto, P.S. 1972. Technology and Cultural Change. Macmillan Publishing Company, New York, New York, USA.

Berry, E.H. 1976. Facilitating rural change. In: Contemporary Africa, edited by C. Gregory and J.F. Neuman. Prentice-Hall, Englewood Cliffs, New Jersey, USA.

Berry, S.S. 1979. Risk and the Poor Farmer. U.S. Agency for International Development, Washington, D.C., USA.

Bhalla, A.J., and Baron, C.G. 1976. Appropriate technology, poverty and unemployment: the ILO Project. Appropriate Technology 1(4):20-22.

Bhatia, R. 1980. Energy and rural development, an analytical framework for socio-economic assessment of technological and policy alternatives. (Available from author, Department of City and Regional Planning, 322 Gund Hall, Harvard University, Cambridge, Massachusetts, USA.)

Bossen, S. 1975. Women in modernizing societies. American Ethnologist 2(4):587-601.

Briscoe, J. 1979. The Political Economy of Energy Use in Rural Bangladesh. Environmental Systems Program, Harvard University, Cambridge, Massachusetts, USA.

Brode, J. 1969. The Process of Modernization: An Annotated Bibliography of the Socio-Cultural Aspects of Development. Harvard University Press, Cambridge, Massachusetts, USA.

Brokensha, D., ed. 1969. The Anthropology of Development in Sub-Saharan Africa. Society for Applied Anthropology Monographs, University of California, Berkeley, California, USA.

Brokensha, D., ed. 1965. Ecology and Economic Development in Tropical Africa. University of California Press, Berkeley, California, USA.

Brown, L. 1980. Food or fuel: new competition for the world's cropland. Worldwatch Paper 35. Worldwatch Institute, Washington, D.C., USA.

Brown, M. 1977. Appropriate Technology for Industry in Developing Countries. Occasional Paper #15. Organisation for Economic Cooperation and Development, Paris, France.

Brown, N.L., ed. 1978. Renewable Energy Resources and Rural Applications in the Developing World. Westview Press, Boulder, Colorado, USA.

Brown, R.H. 1977. Appropriate technology and grass roots: toward a development strategy from the bottom up. Developing Economies 15(3):253-279.

Buvinic, M. 1976. Women and World Development: An Annotated Bibliography. Overseas Development Council, Washington, D.C., USA.

Burrell, G. 1980. Uniform Data Collection and Information Systems for Renewable Energy Projects. Rural Development, Inc., Burlington, Vermont, USA.

Cancian, F. 1968. Stratification and risk taking: a theory tested on agricultural innovation. American Sociological Review 32(6):912-927.

Carr, M. 1978. Economically Appropriate Technologies for Developing Countries: An Annotated Bibliography. Intermediate Technology Publications Ltd., London, England.

Carr, M. 1976. Rural women, rural technology, rural development. Populi 3(4):44-50.

Cecelski, E., Dunkerley, J., and Ramsey, W. 1979. Household Energy and the Poor in the Third World. Resources for the Future, Washington, D.C., USA.

Coburn, B. 1978. Proposal for Appropriate Technology Involvement in the National Parks of Kenya. Threshold, Inc., Washington, D.C., USA.

Cochrane, G. 1971. Development Anthropology. Oxford University Press, New York, New York, USA.

Cohen, R. 1961. The success that failed: an experiment in culture change in Africa. Anthropologica 3:21-36.

Congdon, R.J., ed. 1977. Introduction to Appropriate Technology. Rodale Press, Emmaus, Pennsylvania, USA.

Cook, C.P. n.d. Annotated Bibliography on Energy Technology and Socio-Cultural Change in Africa. Inter-Culture Associates, Columbus, Ohio, USA.

Cook, C.P. n.d. Spontaneous Socio-Technological Change. A Study of Low-Friction Technology Adoption in Africa. Inter-Culture Associates and ATEX, Inc., Columbus, Ohio, USA.

Coombs, P.H., ed. 1980. Meeting the Basic Needs of the Rural Poor: The Integrated Community-Based Approach. A report of the International Council for Education Development. Pergamon Press, New York, New York, USA.

Dalton, G. 1962. Traditional production in primitive African economies. Quarterly Journal of Economics 76(3):360-378.

Dasgupta, S. 1963. Innovation and innovators in Indian villages. Man in India 43:27-34.

Davidson, B. 1974. Can Africa Survive? Arguments Against Growth Without Development. Little, Brown and Company, Boston, Massachusetts, USA.

De Giorgio, A., and Roveda, C., eds. 1980. Criteria for Selecting Appropriate Technologies Under Different Cultural, Technical and Social Conditions. Pergamon Press, New York, New York, USA.

deLucia, R.J. 1980. International Applications of Renewable Energy Resources, Some Comments and Suggestions. Testimony before the U.S. Senate Committee on Energy and National Resources, Subcommittee on Energy Conservation and National Resources. Meta Systems, Inc., Cambridge, Massachusetts, USA.

deLucia, R.J. n.d. A Note on Some Biomass Related Technologies. Meta Systems, Inc., Cambridge, Massachusetts, USA.

deLucia, R.J., and Bhatia, R. 1980. Economics of Renewable Energy Technologies in Rural Third World: A Review. Paper prepared for the Panel on the Introduction and Diffusion of Renewable Energy Technologies. National Academy of Sciences, Washington, D.C., USA.

deLucia, R.J., and Deagle, E.A., Jr. 1981. A Proposal for Energy Planning Assistance to Developing Countries. Meta Systems, Inc., Cambridge, Massachusetts, USA.

Development Alternatives, Inc. 1976. Small Farmer Risk Taking. Report to U.S. Agency for International Development, Washington, D.C., USA.

deWilde, T. 1977. Some social criteria of appropriate technology. In: Introduction to Appropriate Technology, edited by R.J. Congdon. Rodale Press, Emmaus, Pennsylvania, USA.

Diaz Bordenane, J. 1976. Communication of agricultural innovations in Latin America: the need for new models. Communication Research 3:43-62.

Dickinson, H. 1977. Transfer of knowledge and adoption of technologies. In: Introduction to Appropriate Technology, edited by R.J. Congdon. Rodale Press, Emmaus, Pennsylvania, USA.

Diesendorf, M. 1979. Energy and People: Social Implications of Different Energy Futures. Society for Social Responsibility in Science, Canberra, Australia.

Downs, G.W., Jr., and Mohr, L.B. 1976. Conceptual issues in the study of innovation. Administrative Science Quarterly 21:700-714.

Dunkerley, J., Ramsey, W., Gordon, T., and Cecelski, E. 1981. Energy Strategies for Developing Nations. Johns Hopkins University Press, Baltimore, Maryland, USA.

Drucker, C. 1980. Renewable energy technologies for development; soft path to dependency? Soft Energy Notes 3(3):14-15.

Dunn, P.D. 1978. Appropriate Technology: Technology With a Human Face. Schocken, New York, New York, USA.

El-Dayen, A. 1980. Some Sociological Aspects in China. Asian Study Mission Report, Commission on International Relations/Board on Science and Technology for International Development, National Academy of Sciences, Washington, D.C., USA.

Elmendorf, M. 1980. Finding out about energy needs and resources: the human dimension. In: Proceedings. International Workshop on Energy Survey Methodologies for Developing Countries. Jekyll Island, Georgia, 21-25 January. National Academy of Sciences, Washington, D.C., USA.

Ernst, E. 1978. Fuel Consumption among Rural Families in Upper Volta, West Africa. Paper prepared for the 8th World Forestry Congress, 16-27 October 1978, Jakarta, Indonesia.

Esman, M.J. 1974. Popular participation and feedback systems in rural development. In: Communication Strategies for Rural Development, edited by R.H. Crawford and W.B. Ward. Cornell University Press, Ithaca, New York, USA.

Evans, D.D., and Adler, L.N., eds. 1979. Appropriate Technology for Development. Westview Press, Boulder, Colorado, USA.

Evans, I. 1980. A modest proposal to exploit humble foreigners. Rain 7(9):8-9.

Evenson, R. 1974. International diffusion of agrarian technology. Journal of Economic History 34(1):51-73.

Fallen-Bailey, D.G., and Byer, T.A. 1979. Energy Options and Policy Issues in Developing Countries. Staff Working Paper #350. World Bank, Washington, D.C., USA.

Firth, R., and Yamey, B.S. 1964. Capital, Savings, and Credit in Peasant Societies. Chicago University Press, Chicago, Illinois, USA.

Fitzsimmons, A.K., and McIntosh, T.L. 1978. Energy planning in Guatemala. Energy Policy 6:14-20.

Fliegel, F.C. 1968. A cross-national comparison of farmers' perceptions of innovations as related to adaption behavior. Rural Sociology 33:347-449.

Forman, S. 1981. Trip Report to Columbia and Honduras. Prepared for the Panel on the Introduction and Diffusion of Renewable Energy Technologies. National Academy of Sciences, Washington, D.C., USA.

Foster, G.M. 1974. Technical assistance methodology: an anthropologist's view. Development Digest 12(3):87-94.

Franda, M. 1979. India's Rural Development. Indiana University Press, Bloomington, Indiana, USA.

French, D. 1979. The Economics of Renewable Energy Systems for Developing Countries. Report to the al Dir'iyyah Institute and U.S. Agency for International Development, Washington, D.C., USA.

French, D. 1977. Appropriate Technology in Social Context: An Annotated Bibliography. Volunteers in Technical Assistance, Mt. Rainier, Maryland, USA.

Ganiere, N. 1973. Transfer of Technology and Appropriate Technologies: A Bibliography. Organisation for Economic Cooperation and Development, Paris, France.

Goode, P.M. 1975. Village technology for African women. Appropriate Technology 2(3).

Goodman, L.J., and Love, R.N. 1981. Biomass Energy Projects: Planning and Management. Published in cooperation with the East-West Center, Honolulu, Hawaii. Pergamon Press, New York, New York, USA.

Goulet, D. 1977. The Uncertain Promise: Value Conflicts in Technology Transfer. IDOC, North America, Inc., New York, New York, USA, in cooperation with the Overseas Development Council, Washington, D.C., USA.

Goulet, D. 1976. The suppliers and purchasers of technology: a conflict of interests. International Development Review 18(3):14-20.

Griffin, K. 1974. The Political Economy of Agrarian Change: An Essay on the Green Revolution. Harvard University Press, Cambridge, Massachusetts, USA.

Hall, D.O., Barnard, G.W., and Moss, P.A. 1981. Biomass for Energy in the Developing Countries. Pergamon Press, New York, New York, USA.

Hammond, P.B. 1980. Renewable Energy Diffusion in Developing Countries: Towards Strategic Guidelines. Paper prepared for the Panel on the Introduction and Diffusion of Renewable Energy Technologies, National Academy of Sciences, Washington, D.C., USA.

Hammond, P.B. 1978. An Introduction to Cultural and Social Anthropology. Macmillan Publishing Company, New York, New York, USA.

Hammond, P.B. 1978. Social Soundness Analysis. Mali Renewable Energy Project Paper. Center for Research on Economic Development, University of Michigan, Ann Arbor, Michigan, USA.

Hammond, P.B. 1966. Yatenga: Technology in the Culture of a West African Kingdom. Free Press, New York, New York, USA.

Hayes, P. 1982. Social structure and rural energy technology. In: Southern Perspectives on the Rural Energy Crisis, edited by A.K.N. Reddy, R.S. Ganapathy, and P. Hayes. Nautilus, Inc., Bolinas, California, USA.

Hayes, P. 1981. Rural Energy, Technological Change and Competing End-Uses. Prepared for the Panel on the Introduction and Diffusion of Renewable Energy Technologies. National Academy of Sciences, Washington, D.C., USA.

Hoole, P.W., ed. 1979. Evaluation Research and Development Activities. Sage Publications, Inc., Beverly Hills, California, USA.

Hoffman, H.K. 1980. Alternative energy technologies and third world rural energy needs: a case of emerging technological dependence. Development and Change (July).

Howe, J.W. 1980. New village uses of renewable energy sources. In: Village Viability in Contemporary Society. Westview Press, Boulder, Colorado, USA.

Hughart, D. 1979. Prospects for Traditional and Non-conventional Energy Sources in Developing Countries. Staff Working Paper #346. World Bank, Washington, D.C., USA.

Hursh, C.G., and Ray, P. 1976. Third World Surveys: Survey Research in Developing Nations. Macmillan Publishing Co., Delhi, India.

Jackson, S. 1972. Economics of Appropriate Technology for Developing Countries: A Survey. Occasional Paper #3. Overseas Development Council, Washington, D.C., USA.

Jedlicka, A.D. 1977. Organization for Rural Development: Risk Taking and Appropriate Technology. Praeger Publishing Co., New York, New York, USA.

Jedlicka, A.D. 1975. Diffusion of Technical Innovation: A Case for the Non-Sexist Approach among Rural Villages. Paper prepared for a Seminar on Women in Development sponsored by the American Association for the Advancement of Science, Mexico City, Mexico.

Jequier, M., ed. 1976. Appropriate Technology: Problems and Promises. Organisation for Economic Cooperation and Development, Paris, France.

Jequier, M. 1975. Intermediate technology: a new approach to development problems. OECD Observer 75:26-28.

Johnston, P. 1974. Appropriate Technologies for Small Developing Countries. Smoothie Publishing Company, Brighton, England.

Jones, G.E. 1967. The adoption and diffusion of agricultural practices. World Agricultural Economics and Rural Sociology Abstracts 9:1-34.

Kelly, P., and Kranzberg, M. 1975. Technological Innovation: A Critical Review of Current Knowledge. Georgia Institute of Technology, Atlanta, Georgia, USA.

Khan, A.U. 1974. Appropriate technologies: do we transfer, adopt or develop? In: Employment in Developing Nations: A Report on a Ford Foundation Study, edited by E.O. Edwards. Columbia University Press, New York, New York, USA.

Knight, C.G., and Neuman, J.G., eds. 1976. Contemporary Africa--Geography and Change. Prentice-Hall, Inc., Englewood Cliffs, New Jersey, USA.

Lance, L.M., and McKenna, E.E. 1975. Analysis of cases pertaining to the impact of Western technology on the non-Western world. Human Organization 34(1):87-94.

Lear, J.G. 1974. Rural Development: Recognizing the Role of Women, A Selected Bibliography. National Planning Association, Washington, D.C., USA.

Lester, J.M. 1974. Technology Transfer and Developing Countries: A Selected Bibliography. George Washington University Press, Washington, D.C., USA.

Lipinsky, E.S. 1978. Fuels from biomass: integration with food and material systems. Science 199:644-651.

Lipton, M. 1977. Let the Poor Stay Poor: Urban Bias in World Development. Harvard University Press, Cambridge, Massachusetts, USA.

MacPherson, G., and Jackson, D. 1975. Village technology for rural development: agricultural innovation in Tanzania. International Labor Review 3(2):102-112.

Makhijani, A. 1978. Economics and Sociology of Alternate Energy Sources. UNDP seminar paper FP/0404-78-04(C1092).

Makhijani, A. 1976. Energy Policy for the Rural Third World. International Institute for Environment and Development, Washington, D.C., USA.

Makhijani, A. 1976. Solar Energy and Rural Development for the Third World. Ballinger, Cambridge, Massachusetts, USA.

Malawi Energy Unit. 1981. Energy Notes 1(2):1-2. Ministry of Agriculture, Box 30134, Lilongwe 3, Malawi.

Mann, R.D. 1976. A Survey Technique for Identifying the Needs of Small Farmers, and an Example of its Use in Zambia. Intermediate Technology Publishers, London, England.

Masrui, A. 1972. Cultural engineering and socialized technology. In: Cultural Engineering and Culture Building in East Africa, edited by A. Masrui. Northwestern University Press, Evanston, Illinois, USA.

Meadows, D.H., Meadows, D.S., Randers, J., and Bebrans, W.W. 1972. The Limits of Land. Universe Books, New York, New York, USA.

Melchor, A., Jr. 1981. Options Immediately Available to the Region's Developing Countries to Lessen their Dependence on Imported Hydrocarbon Fuels. United Nations University, Tokyo, Japan.

Misch, M.R., and Margolin, J.B. 1975. Rural Women's Groups as Potential Change Agents: A Study of Colombia, Korea, and the Philippines. George Washington University Program of Policy Studies in Science and Technology, George Washington University, Washington, D.C., USA.

Morrison, D.E. 1978. Energy, Appropriate Technology and International Interdependence. Paper prepared for the Annual Meeting of the Society for the Study of Social Problems, San Francisco, California, USA.

Morse, R., and Fesharaki, F. 1980. Assessing Alternative Resources, Technologies, and Organizational Means for Meeting Energy Needs. Energy for Rural Development Program Report, PR-80-1. East-West Center, Honolulu, Hawaii, USA.

McInerney, J.P. 1978. The Technology of Rural Development. Staff Working Paper #295. World Bank, Washington, D.C., USA.

Nash, H. 1977. Progress as if Survival Mattered. Friends of the Earth, San Francisco, California, USA.

National Academy of Sciences. 1981. Staff Report: Environmental Degradation in Mauritania. National Academy Press, Washington, D.C., USA.

National Academy of Sciences. 1981. Workshop on Energy and Agriculture in Developing Countries. National Academy Press, Washington, D.C., USA.

National Academy of Sciences. 1980. Proceedings. International Workshop on Energy Survey Methodologies for Developing Countries. National Academy Press, Washington, D.C., USA.

National Academy of Sciences. 1980. Firewood Crops. National Academy Press, Washington, D.C., USA.

National Academy of Sciences. 1977. Appropriate Technologies for Developing Countries. National Academy of Sciences, Washington, D.C., USA.

National Academy of Sciences. 1976. Energy for Rural Development: Renewable Resources and Alternative Technologies for Developing Countries. National Academy of Sciences, Washington, D.C., USA.

Noronha, R. 1980. Sociological Aspects of Forestry Project Design. Department of Agriculture and Rural Development. World Bank, Washington, D.C., USA.

Obibuaku, S.O. 1979. Technical assistance and agricultural change in some Anambra state villages, Nigeria. Journal of Developing Areas 14:43-53.

Organisation for Economic Cooperation and Development. 1975. Low Cost Technology: An Inquiry into Outstanding Policy Issues. OECD, Paris, France.

O'Kelly, E. n.d. Appropriate technology for women of the developing countries. Peace Corps Program and Training Journal 4(6):10-13.

Pak, S.J., and Taylor, C.R. 1976. Critical Factors in Economic Evaluation of Small Decentralized Energy Projects. Science and Technology Report #25. World Bank, Washington, D.C., USA.

Pala, A.O. 1975. The Role of African Women in Rural Development: Research Priorities. Institute for Development Studies Discussion Paper #203. University of Nairobi, Nairobi, Kenya.

Perkins, D., et al. 1977. Rural Small-scale Industry in the People's Republic of China. California University Press, Berkeley, California, USA.

Practical Concepts Incorporated. 1980. Planning Rural Energy Projects. A Rural Energy Survey and Planning Methodology for Bolivia. Available from PCI, 1730 Rhode Island Avenue, N.W., Washington, D.C., USA.

Prakasam, T.B.S. 1981. Some Aspects of Biogas Technology in India and Egypt. Paper prepared for the Panel on the Introduction and Diffusion of Renewable Energy Technologies. National Academy of Sciences, Washington, D.C., USA.

Radnor, M., Feller, I., and Rogers, E.M., eds. 1978. The Diffusion of Innovations: An Assessment. Center for the Inter-Disciplinary Study of Science and Technology, Evanston, Illinois, USA.

Raghanen, S. 1974. Women's Role and Development Policies: A Bibliographic Index. Society for International Development, Committee on Women in Development, Washington, D.C., USA.

Rao, R. 1980. When alternatives are inappropriate. New Scientist 28(30):4.

Reddy, A.K.N. 1979. Alternative and Traditional Energy Sources for the Rural Areas of the Third World. Paper prepared for the Royal Institute Forum on Third World Energy Strategies and the Role of Industrialized Countries, 20-27 June, London, England.

Reddy, A.K.N. 1978. Methodology for the selection of environmentally sound and appropriate technologies. Soft Energy Notes 1:59-60.

Revelle, R. 1979. Energy Sources for Rural Development. Paper presented at the Conference on Energy Alternatives of the United Nations University, 9-12 January. East-West Center, Honolulu, Hawaii, USA.

Riegelman, M.A. 1975. A Seven Country Survey on the Roles of Women in Rural Development. Development Alternatives, Inc., Washington, D.C., USA.

Robinson, A., ed. 1979. Appropriate Technologies for Third World Development. St. Martins Press, New York, New York, USA.

Rochin, R.I. 1978. Why farmers do not adopt and utilize new technology. In: Proceedings and Issues of the CENTO Seminar on Increasing the Productive Capacity of Small Farmers in CENTO Countries, 17-21 December, Lahore, Pakistan. Central Treaty Organization, Ankara, Turkey.

Roeling, N.G., et al. 1976. The diffusion of innovations and the issue of equity in rural development. Communication Research 3:63-78.

Rogers, E.M. 1980. The Diffusion of Technological Innovations: Application to Renewable Energy Resources in Developing Nations. Paper prepared for the Panel on the Introduction and Diffusion of Renewable Energy Technologies. National Academy of Sciences, Washington, D.C., USA.

Rogers, E.M., ed. 1976. Communication and Development: Critical Perspectives. Sage, Beverly Hills, California, USA.

Rogers, E.M. 1976. The passing of the dominant paradigm: reflections on diffusion research. In: Communication and Change: The Last Ten Years and the Next, edited by W. Schramm and D. Lerner. East-West Center, Honolulu, Hawaii, USA.

Rogers, E.M. 1976. Where are we in understanding the diffusion of innovations? In: Communication and Change: The Last Ten Years and the Next, edited by W. Schramm and D. Lerner. East-West Center, Honolulu, Hawaii, USA.

Rogers, E.M. 1964. Bibliography of Research on the Diffusion of Innovations. Department of Communications, Michigan State University, East Lansing, Michigan, USA.

Rogers, E.M. 1962. Diffusion of Innovations. Free Press, New York, New York, USA.

Rothman, J. 1974. Planning and Organizing for Social Change: Action Principles for Social Science Research. Columbia University Press, New York, New York, USA.

Rounaq, J. 1975. Women in Bangladesh. In: Women Cross-Culturally: Change and Challenge, edited by R. Rohlich-Leavitt. Mouton World Anthropology Series, Mouton, The Hague, The Netherlands.

Ruttan, V.W., and Hayami, Y. 1973. Technology transfer in agricultural development. Technology and Culture 14:119-151.

Schlegel, C.C., and Tarrant, J. 1980. Thinking About Energy and Rural Development: Methodological Guidelines for Socioeconomic Assessment. East-West Center, Honolulu, Hawaii, USA.

Schutjer, W.A., and Van Der Veen, M.G. 1976. Economic Constraints on Agricultural Technology Adoption in Developing Nations. Report available from the Department of Agricultural Economics and Rural Sociology, Agricultural Experiment Station, Pennsylvania State University, University Park, Pennsylvania, USA.

Scott, J. 1976. The Moral Economy of the Peasant. Yale University Press, New Haven, Connecticut, USA.

Self, G. 1979. Design of Community Renewable Energy Projects. U.S. Agency for International Development, Washington, D.C., USA.

Shell Briefing Service. 1980. Energy from Biomass. Newgate, London, England.

Sigurdson, J. 1979. Rural Industrialization in China. Harvard University Press, Cambridge, Massachusetts, USA.

Singer, H. 1977. Technologies for Basic Needs. International Labour Office, Geneva, Switzerland.

Smil, V. 1981. Energy in General and on Biomass in Particular. Paper prepared for the Panel on the Introduction and Diffusion of Renewable Energy Technologies. National Academy of Sciences, Washington, D.C., USA.

Smil, V. 1979. Energy flows in the developing world. American Scientist 67(5):522-531.

Smith, K., Santerre, M., and Schlegel, S. 1980. Criterion Framework and Indicators for Comparing and Evaluating Alternative Energy Technologies. Resource Systems Institute Conference Paper. East-West Center, Honolulu, Hawaii, USA.

Smith, K., and Santerre, M. 1980. Criteria for Evaluating Small-scale Rural Energy Technologies: The Flert Approach. East-West Center, Honolulu, Hawaii, USA.

Spencer, D.S., and Woroniak, A., eds. 1967. The Transfer of Technology to Developing Countries. Praeger Publishing Co., New York, New York, USA.

Stewart, H. 1979. Technology and Underdevelopment. Westview Press, Boulder, Colorado, USA.

Taylor, J. 1979. From Modernization to Modes of Production: A Critique of the Sociologies of Development and Underdevelopment. Humanities Press, Atlantic Highlands, New Jersey, USA.

Taylor, K.S. 1978. Sugar and the Underdevelopment of Northeastern Brazil. University of Florida Press, Gainesville, Florida, USA.

Thomas, J. 1975. The Choice of Technology in Developing Countries. Harvard University Press, Cambridge, Massachusetts, USA.

Tinker, I., and Bramsen, M.B., eds. 1976. Women and World Development. American Association for the Advancement of Science/Overseas Development Council, Washington, D.C., USA.

United Nations Development Programme. 1980. Rural Women's Participation in Development. Evaluation Study #3. UNDP, New York, New York, USA.

United Nations Economic Commission for Africa. 1972. Women: the neglected human resource for African development. Canadian Journal of African Studies 16(2):359-370.

Van Allen, J. 1974. Women in Africa, modernization means more dependency. The Center Magazine 60-67.

Villegas, B.M. 1974. Korean appropriate technology that uses the resources of the third world and substitutes labor for capital. Ceres 7(3):44-47.

Von Blankenburg, P., and Schultz, M. 1970. The socio-economic context of agricultural innovation processes. Landwirtschaft 9(4):317-333.

Walsh, J. 1981. Technology transfer reappraised. Science 21:902.

Walton, J.D., Roy, A.H., and Bomar, S.H., Jr. 1978. A State of the Art Survey of Solar Powered Irrigation Pumps, Solar Cookers, and Wood Burning Stoves for Use in Sub-Saharan Africa. Georgia Institute of Technology, Atlanta, Georgia, USA.

Ward, B.E. 1970. Women and technology in developing countries. Impact of Science on Society 20(1):93-101.

Warner, K. 1974. The need for some innovative concepts of innovation. Policy Sciences 5:443-451.

Weiss, C., Jr. 1979. Mobilizing technology for developing countries. Science 203:1083-1089.

Wharton, C.R., Jr. 1968. Risk, Uncertainty and the Subsistence Farmer: Techno-Ecological Innovation and Resistance to Change in the Context of Survival. Paper presented at the Joint Session of the American Economic Association and the Association for Comparative Economics, 28 December, Chicago, Ilinois, USA.

Wilkes, C.K. ed. 1973. The Political Economy of Development and Underdevelopment. Random House, New York, New York, USA.

Wolfe, M. 1975. Participation of women in development in Latin America. In: Women in Latin America: Three Contributions to a Discussion, edited by M. Wolfe, J. Garciarena, and H. Kirsch. United Nations Development Programme, 29 April-2 May, Caracas, Venezuela.

World Bank. 1982. Renewable Energy Sources. World Bank, Washington, D.C., USA.

World Bank. 1980. Energy in the Developing Countries. World Bank, Washington, D.C., USA.

World Bank. 1978. Appropriate Technology and World Bank Assistance to the Poor. World Bank, Washington, D.C., USA.

World Bank. 1976. Appropriate Technology in World Bank Activities. World Bank, Washington, D.C., USA.

Wright, E.H. 1979. Introducing new technologies for alternative energy sources: the economic, social, and environmental considerations for developing countries. Energy Economics and Policy:4919-4925.

Wunsch, J.S. 1980. Renewable Resource Management. Decentralization and Localization in the Sahel: The Case of Afforestation. Manuscript.

AGROFORESTRY/SOCIAL FORESTRY/WOODLOTS

Ahn, B.W. 1978. Village Forestry in Korea. Paper presented at the 8th World Forestry Congress, 16-27 October, 1978, Jakarta, Indonesia.

Arnold, J.E.M., et al. 1979. Economic Analysis of Forestry Projects: Case Studies Forestry Paper 17, Supplement. Local Community Development. Food and Agriculture Organization, Rome, Italy.

Barnett, A. 1980. The Factors Affecting the Diffusion of Wood Fuel Technologies in the Rural Third World. Science Policy Research Unit, University of Sussex, Brighton, England.

Bene, J.G., Beale, H.W., and Cote, A. 1977. Trees, Food and People: Land Management in the Tropics. International Development Research Centre, Ottawa, Canada.

Bhaja, S.K. 1977. Social forestry in Tamil Nadu. Indian Farming 26(11):8.

Brokensha, D. and Riley, B. 1978. Forestry, Foraging, Fences and Fuel in a Marginal Area of Kenya. Paper prepared for Africa Bureau Firewood Workshop, 12-14 June. U.S. Agency for International Development, Washington, D.C., USA.

Castanos, M., and Leon, J. 1978. Implementacion de Programas Forestales en Mèxico para el Desarrollo de las Communidades Locales. Paper presented at the Eighth World Forestry Conference, Jakarta, 16-27 October, 1978. Food and Agriculture Organization of the United Nations, Rome, Italy.

Cernea, M. 1980. Land Tenure Systems and Social Implications of a Forestry Development Program--a Case Study in Azad Kashmir. World Bank, Washington, D.C., USA.

Chandler, T., and Spurgeon, D., eds. 1979. International Cooperation in Agroforestry: Proceedings of an International Conference. Sponsored by German Foundation for International Development, Federal Republic of Germany and International Council for Research in Agroforestry, Nairobi, Kenya.

Digernes, T.H. 1977. Wood for Fuel--Energy Crisis Implying Desertification: The Case of Bara, the Sudan. Ph.D. Thesis, Department of Geography, University of Bergen, Norway.

Done, M. 1980. Development of tribal land rights in Borneo: the role of ecological factors. Borneo Research Bulletin 12(1):3-19.

Eckholm, E. 1979. Planting for the future: forestry for human needs. Worldwatch Paper 26. Worldwatch Institute, Washington, D.C., USA.

Eckholm, E. 1975. The other energy crisis: firewood. Worldwatch Paper 1. Worldwatch Institute, Washington, D.C., USA.

Fleuret, P.C., and Fleuret, A. 1978. Fuelwood use in a peasant community: a Tanzanian case study. The Journal of Developing Areas 12:315-322.

Food and Agriculture Organization of the United Nations. 1981. Map of the Fuelwood Situation in Developing Countries. FAO, Rome, Italy.

Freeman, P.H. 1979. Forestry in Development Assistance: Background Paper for AID Policy and Programs. U.S. Agency For International Development, Development Support Bureau, Washington, D.C., USA.

French, D. 1978. Firewood in Africa. Paper prepared for Africa Bureau Firewood Workshop, 12-14 June. U.S. Agency for International Development, Washington, D.C., USA.

Gamser, M.S. 1980. The forest resource and rural energy development. World Development 8:769-780.

Gregersen, H.M. 1982. Village Forestry Development in the Republic of Korea: A Case Study. Food and Agriculture Organization of the United Nations, Rome, Italy.

Gulick, F. 1979. Fuelwood and Other Renewable Energies in Africa: A Working Inventory of Current Programs and Proposals for ODC-AID. Paper prepared for Africa Bureau Firewood Workshop, 12-14 June. U.S. Agency for International Development, Washington, D.C., USA.

Hall, D.O., and Coombs, J. 1981. Biomass Production in Agroforestry for Fuels and Food. Prepared for ICRAF Consultative Meeting on Plant Research and Agroforestry. International Council for Research in Agroforestry, Nairobi, Kenya.

Hammond, P.B. 1980. Issues in Social Forestry. Manuscript, 22 April. World Bank, Washington, D.C., USA.

Hoskins, M. 1981. Communication and energy: community participation in forestry projects. Development Communication Report No. 33.

Hoskins, M. 1980. Social Dimensions in Local Forestry/Conservation Efforts, Manuscript. World Bank, Washington, D.C., USA.

Hoskins, M. 1979. Community Participation in African Fuelwood Production, Transformation and Utilization. Report to Overseas Development Council and U.S. Agency for International Development, Washington, D.C., USA.

Hoskins, M. 1979. Women in Forestry for Local Community Development, A Programming Guide. Office of Women in Development, U.S. Agency for International Development, Washington, D.C., USA.

Iyamabo, D.E., Onweluzo, B.S.K., and Okrie, P.E. 1977. Farm Forestry in Nigeria: Case Study on Forestry for Community Development. Food and Agriculture Organization of the United Nations, Rome, Italy.

Karamchandani, K.P. 1981. Gujarat Social Forestry: A Case Study. Food and Agriculture Organization of the United Nations, Rome, Italy.

Keita, J.D. 1979. Elements de Strategie pour la Satisfaction des Besoins en Combustibles Ligneux du Sahel d'Ici l'An 2000. CILSS/Club du Sahel, Niamey, Niger.

Ki-Zerbo, J. 1980. The shortage of firewood is a daily vital concern. VITA News, Special Energy Issue 15 July.

Maghembe, J.A., and Redhead, J.F. 1980. Agroforestry: preliminary results of intercropping acacia, eucalyptus and leucaena with maize and beans. Division of Forestry, University of Dar-es-Salaam, Morogoro, Tanzania.

Moss, R.P., and Morgan, W.P. 1981. Fuelwood and Rural Energy Production and Supply in the Humid Tropics. United Nations University, Tokyo, Japan.

National Research Council. 1983. Agroforestry in the West African Sahel. National Academy Press, Washington, D.C., USA.

National Research Council. 1983. Environmental Change in the West African Sahel. National Academy Press, Washington, D.C., USA.

Noronha, R. 1981. Why is it so difficult to grow fuelwood? Unasylva 33(13).

Noronha, R. 1980. Village Woodlots: Are They a Solution? Paper prepared for the Panel on the Introduction and Diffusion of Renewable Energy Technologies. National Academy of Sciences, Washington, D.C., USA.

Poulsen, G. 1979. Woodlots in the Sahel--is that the solution? Sylva Africana 4:3-4.

Poulsen, G. 1978. Man and Tree and Tropical Africa. International Development Research Centre, Ottawa, Canada.

Richardson, S.D. 1966. Forestry in Communist China. Johns Hopkins University Press, Baltimore, Maryland, USA.

Salem, B., and Tran van Nao. 1980. Fuelwood Production in Traditional Farming System. Food and Agriculture Organization of the United Nations, Rome, Italy.

Spears, J.S. 1978. The Changing Emphasis in World Bank Lending: A Summary of Recent Experiences and Problem Areas of Relevance. Paper presented at the Eighth World Forestry Congress, Jakarta, Indonesia, 16-27 October 1978. Food and Agriculture Organization of the United Nations, Rome, Italy.

Spears, J.S. 1978. Wood as an Energy Source: The Situation in the Developing World. Paper presented at the 103rd Annual Meeting of the American Forestry Association, Hot Springs, Arkansas, USA.

Spurgeon, D. 1980. Agroforestry: a promising system of improved land management in Latin American. Interciencia 5(3):176-178.

Thomson, J.T. 1980. Firewood Survey: Theory and Methodology. U.S. Agency for International Development, Washington, D.C., USA.

Thomson, J.T. 1980. Trouble Case Investigation of Problems in Nigerian Rural Modernization: Forest Conservation. Manuscript, available from the author, Department of Government, Lafayette College, Easton, Pennsylvania, USA.

Tisseverasinghe, A.E.K. 1980. Agro-forestry for Papua New Guinea. Journal of the Forestry Society of the Papua New Guinea University of Technology 1(4):37-49.

Ulinski, C.A. 1978. U.S. Forestry and Ecology Program in the Sahel. U.S. Agency for International Development, Washington, D.C., USA.

Umali, D.C. 1979. Keynote Address to FAO/SIDA Seminar on Forestry in Rural Community Development, 13-15 December 1979, Chiang Mai, Thailand.

Weber, F.R. 1977. Reforestation in Arid Lands. Volunteers in Technical Assistance, Mt. Rainier, Maryland, USA.

Westoby, J.C. 1978. Forest industries for socio-economic development. Commonwealth Forestry Review 58(2):107-116.

Winterbottom, R. 1979. Upper Volta Koudougou Agricultural Development Project. Appraisal Report for the Forestry Sub-Program. West Africa Projects Department, World Bank, Washington, D.C., USA.

World Bank. 1978. Forestry Sector Policy Paper. World Bank, Washington, D.C., USA.

Wood, D., et al. 1979. The Socio-Economic and Environmental Context of Fuelwood Use in Rural Communities of Developing Countries: Issues and Guidelines for Community Fuelwood Packages. U.S. Agency for International Development, Washington, D.C., USA.

CHARCOAL

Bertrand, A. 1977. Les problèmes du bois de chauffage et du carbon du bois en Afrique tropicale. Revue de Bois et Forêts des Tropiques 173:39-48.

Blowers, M.J. 1980. Small scale charcoal production using the West Indian and Tongan type kilns. Journal of the Forestry Society of the Papua New Guinea University of Technology 1(4):23-36.

Earl, D.E. 1974. Charcoal: an Andre Mayer Fellowship Report. Food and Agriculture Organization of the United Nations, Rome, Italy.

Earl, D.E., and Earl, A. 1975. Charcoal Making for Small Scale Enterprises: An Illustrated Training Manual. International Labour Organization, Geneva, Switzerland.

Food and Agriculture Organization of the United Nations. 1979. Issue Paper for the Technical Panel on Fuelwood and Charcoal. Prepared for the Preparatory Committee's Technical Panel for the United Nations Conference on New and Renewable Sources of Energy, 21-25 January 1980. FAO, Rome, Italy.

Garriott, G. 1982. Four improved charcoal kiln designs. VITA Energy Bulletin 2(1):11-13,15.

Hehr, J.J. 1967. Charcoal: Its Multifarious Effects in a Rural Guatemalan Community. M.A. Thesis, Michigan State University, East Lansing, Michigan, USA.

Ki-Zerbo, J., and De Lepeleire, G. 1979. L'Amelioration des Foyers pour l'Utilisation Domestique du Bois de Feu: Ses Possibilites et Son Impact au Sahel. Comité Inter-état pour la Lutte contre la Sécheresse au Sahel and Club du Sahel, Niamey, Niger.

Little, E.C.S. 1978. The mini-cusab kiln for rapid small-scale manufacture of charcoal from scrub, coconut wood, and coconut shells. Appropriate Technology 5(1):12-14.

Little, E.C.S. 1972. A kiln for charcoal making in the field. Tropical Science 14(3):261-270.

Martin, L.H. 1977. The Ecology and Economics of Cooking Fuels in Ghana. M.A. Thesis, American University, Washington, D.C., USA.

Milhayi, L. 1972. On the Investigation Into Nature, Economic and Social Significance of the Charcoal Industry in Zambia. Department of Geography, University of Zambia.

Winslow, A. 1979. Charcoal (Annotated Bibliography). Commonwealth Forestry Bureau, Oxford, England.

GASIFICATION

Breag, G.R., and Chittenden, A.E. 1979. Producer Gas: Its Potential and Application in Developing Countries. Tropical Products Institute, London, England. (Publication G-130.)

Doroteo, N.B., and Carrillo, V.H. 1980. Producer gas: energy from rice husk, charcoal, wood. Pipeline 1(1):4-7.

Eoff, K.M., and Post, F.M. 1980. How to Power a Gasoline Engine with Wood. Institute of Food and Agricultural Sciences, University of Florida, Gainesville, Florida, USA.

Goss, J.R., and Coppick, R.H. 1980. Producing gas from crop residues. California Agriculture 34(5):4-6.

Groenveld, M.J., and Westerterp, K.R. 1980. Social and economical aspects of the introduction of gasification technology in the rural areas in developing countries. Chemical Age of India 31:185-193.

Onaji, P.B., Adelfila, S.S., and Beenackers, A. 1980. Economic feasibility of gasification in Nigeria. Chemical Age of India 31:194-197.

Rogers, K.E., ed. 1979. Technology and Economics of Wood Residue Gasification. Proceedings of the Tenth Texas Industrial Wood Seminar, 13 March 1979, Lufkin, Texas, USA.

COOKSTOVES

Anonymous. 1980. Des cuisinières à bois et à charbon pour la survie du Sahel. Le Soleil 4 August.

Anonymous. 1979. The Tungko earth stove. Earthman 3-4 December.

Brown, J. 1977. Technical Summary of the Lime Kiln Project in Honduras. Volunteers in Technical Assistance, Mt. Rainier, Maryland, USA.

Centro de Desarrollo Industrial. Programa de Tecnologias Rurales. 1980. La Estufa Lorena. Servicio de Informacion y Recursos Técnicos, Tegucigalpa, Honduras.

Corillon, C., and Gritzner, J. 1981. Discussions in the West African Sahel concerning renewable energy technologies, with particular emphasis upon efforts to evaluate and introduce more efficient wood-burning stoves. National Academy of Sciences, Washington, D.C., USA.

deSilva, D. 1981. A charcoal stove from Sri Lanka. Appropriate Technology 7(4):22-24.

Dutt, G.S. 1981. Field Evaluation of Wood Stoves (with Special Reference to West Africa). Volunteers in Technical Assistance, Mt. Rainier, Maryland, USA.

Dutt, G.S. n.d. Efficient wood burning cooking stove literature. Center for Environmental Studies, Princeton University, Princeton, New Jersey, USA.

Edwards, S.S. 1979. Central America: the Lorena cookstove. In: Appropriate Technology for Development, edited by D. Evans, and L. Adler. Westview Press, Boulder, Colorado, USA.

Elmendorf, C.L. 1980. Case Study on Fechovil's Program of Improved Stove Diffusion in Honduras. Foundation for Cooperative Housing, Washington, D.C., USA.

Evans, I. 1979. *Lorena Owner-Built Stoves*. Volunteers in Asia, Stanford, California, USA.

Evans, I., Gern, E., and Jacobs, L. 1980. *Improved Cookstoves for Rural Senegal*. Volunteers in Technical Assistance, Mt. Rainier, Maryland, USA.

Geller, H.S. 1981. Comparison of Efficiency Between Traditional and Hyderabad Cookstoves. Centre for the Application of Science and Technology to Rural Areas (ASTRA), Indian Institute of Science, Bangalore, India.

Geller, H.S. 1981. Cooking in the Ungra Area: Fuel Efficiency, Energy Losses, and Opportunities for Reducing Firewood Consumption. Centre for the Application of Science and Technology to Rural Areas (ASTRA), Indian Institute of Science, Bangalore, India.

Geller, H.S. 1981. Experiments with Modified and New Woodburning Cookstoves--Efficiency Test Results. Centre for the Application on Science and Technology to Rural Areas (ASTRA), Indian Institute of Science, Bangalore, India.

German Appropriate Technology Exchange. 1980. *Helping People in Poor Countries Develop Fuel-Saving Cookstoves*. Aprovecho Institute, Eugene, Oregon, USA.

Instituto Centroamericano de Investigacion y Tecnologia Industrial. 1981. *Guia Para la Construccion, Uso y Mantenimiento de Estufas Doméstícas de Lena*. ICAITI, Guatemala.

Joseph, S., and Shanahan, Y. 1981. *Laboratory and Field Testing of Monolithic Mud Stoves*. Intermediate Technology Development Group, Ltd., London, England.

Joseph, S., and Trussell, J. 1981. Report on Advisory Visit to the VITA Wood Stove Project in Upper Volta, 16 February to 2 March, 1981. Prepared for Volunteers in Technical Assistance. Intermediate Technology Consultants, Ltd., London, United Kingdom.

Ki-Zerbo, J. 1980. *Improved Woodstoves: User's Needs and Expectations in Upper Volta*. Volunteers in Technical Assistance, Mt. Rainier, Maryland, USA.

Ki-Zerbo, J., and De Lepeleire, G. 1979. L'Amelioration des Foyers pour l'Utilisation Domestique du Bois de Feu: Ses Possibilités et Son Impact au Sahel. Comité Inter-état pour la Lutte contre la Sécheresse au Sahel and Club du Sahel, Niamey, Niger.

Knowland, W. 1980. Better stoves and forests in Indonesia. Open letter to Peter B. Martin, Executive Director, Institute of Current World Affairs, Wheelock House, Hanover, New Hampshire, USA.

Micuta, W. 1981. Modern Stoves for All. Fondation de Bellerive, Geneva, Switzerland.

Ministerio de Agricultura. Instituto Colombiano Agropecuario. n.d. Estufa de Carbon (o lena). Servicio Nacional de Planos, Seccion de Infraestructura, Colombia.

Norman, H.R. 1980. Proposed Project to Develop Woodstoves in Upper Volta. Volunteers in Technical Assistance, Mt. Rainier, Maryland, USA.

Prasad, K.K., ed. 1981. A Study on the Performance of Two Metal Stoves. Woodburning Stove Group, Eindhoven University of Technology and Divisions of Technology for Society, TNO, Apeldoorn, The Netherlands.

Roggeman, J.B. 1980. Les Fourneaux Ameliorès dans le Sahel. Organisation for Economic Cooperation and Development, Paris, France.

Singer, H. 1961. Report to the Government of Indonesia on Improvement of Fuelwood Cooking Stoves and Economy in Fuel Wood Consumption. Food and Agriculture Organization of the United Nations, Rome, Italy.

South Pacific Commission. Community Education Training Centre. n.d. How to Build a Smokeless Stove. South Pacific Commission, Noumea Cedex, New Caledonia.

Strehlke, B. 1979. Intermediate technology for forestry in developing countries. Forstarchiv 40(9):203-204.

Volunteers in Technical Assistance. 1980. Wood Conserving Cookstoves: A Design Guide. VITA, Mt. Rainier, Maryland, USA.

Walton, J.D., Jr., Roy, A.H., and Bomar, S.H., Jr. 1978. A State of the Art Survey of Solar Powered Irrigation Pumps, Solar Cookers, and Wood Burning Stoves for Use in Sub-Saharan Africa. Georgia Institute of Technology, Atlanta, Georgia, USA.

Weir, A., and Richolson, J. 1980. The Nasinu Wood-Burning Stoves: A Preliminary Report. Fiji Forestry Department/University of the South Pacific, Suva, Fiji.

Wood, T.S. 1981. Laboratory and Field Testing of Improved Woodstoves in Upper Volta. Paper prepared for Panel on the Introduction and Diffusion of Renewable Energy Technologies. National Academy of Sciences, Washington, D.C., USA.

BIOGAS

Barnett, A., et al. 1978. Biogas Technology in the Third World. International Development Research Centre, Ottawa, Canada.

Bhatia, R. 1977. Economic Appraisal of Biogas Units in India: A Framework for Social Benefit Cost Analysis. Institute for Economic Growth, New Delhi, India.

Fuller, R. 1980. Bangladesh bio-gas developments. ADAB News (Dacca) 3(7):8-9.

Ghate, P.B. 1979. Biogas: a decentralized energy system. Economic and Political Weekly (Bombay), India, July 7:1132-1136.

Khandelwal, K.C. 1979. Implementation of biogas programme at village level--a case study. Abstracts, National Symposium on Biogas Technology and Usage. Indian Agricultural Research Institute, New Delhi, India.

Loehr, R.D. 1978. Methane from human, animal and agricultural wastes. In: Renewable Energy Resources and Rural Applications in the Developing World, edited by N.L. Brown. Westview Press, Boulder, Colorado, USA.

McGarry, M.G., and Stanforth, J., eds. 1978. Compost, Fertilizer, and Biogas Production from Human and Farm Wastes, in the People's Republic of China. International Development Research Centre, Ottawa, Canada.

National Academy of Sciences. 1979. Methane Generation from Human, Animal, and Agricultural Wastes. National Academy of Sciences, Washington, D.C., USA.

Openshaw, K. 1980. The fuel problem grows larger as African forests get smaller. VITA News July:11-13.

Roberts, R.S., Jr. 1979. Tanzania: biogas generator. In: Appropriate Technology for Development, edited by D.D. Evans and L.N. Adler. Westview Press, Boulder, Colorado, USA.

Ray, P. 1977. Social Aspects of Biogas Plants. Paper presented at the Workshop on Bio-Gas Systems, Management Development Institute, New Delhi, India.

Sathianathan, M.A. 1975. Bio-gas Achievements and Challenges. Association of Voluntary Agencies for Rural Development, New Delhi, India.

Singh, K.H. 1979. Extension aspects of cow dung gas plant. Abstracts, National Symposium of Biogas Technology and Usage, New Delhi, 29-30 November. Indian Agricultural Research Institute, New Delhi, India.

Surahmanian, S.K. 1977. Bio-gas Systems in Asia. Management Development Institute, New Delhi, India.

Van Buren, A., ed. 1976. A Chinese Biogas Manual. Intermediate Technology Publications Limited, London, England.

ALCOHOL FUELS

Adams, R.I. 1979. Agricultural Adjustments to Brazil's Alcohol Program: A Regional Economic Analysis. Ohio State University, Columbus, Ohio, USA.

Anonymous. 1980. Ethanol project to be set up to improve the income of farmers. Indonesian Observer September 12.

Bhatia, R.K. 1980. Fuel Alcohol from Agro Products: A Conceptual Framework for a Study in India and Papua New Guinea. Department of City and Regional Planning, Harvard University, Cambridge, Massachusetts, USA.

Carmo, A.T. 1978. O Programa Nacional do Alcool: Balanços e Perspectivos. Brasil Acucareiro, 17-24 October.

Chancellor, W.J. 1978. The role of fuel and electrical energy in increasing production for traditionally based agriculture. Technical Paper of the American Society of Agricultural Engineers 78-5529. American Society of Agricultural Engineers, St. Joseph, Michigan, USA.

DaSilva, J.G., et al. 1978. Energy balance for ethyl alcohol production for crops. Science 201:903-906.

Hammond, A.L. 1977. Alcohol: a Brazilian answer to the energy crisis. Science 195:566-567.

Jones, J.L., et al. 1978. Alcohol Fuels Production Technology and Economics. Stanford Research Institute, Menlo Park, California, USA.

Kuhner, J. 1980. Impact of biomass-derived energy programs. In: IV International Symposium on Alcohol Fuels Technology, 5-8 October, Guaruja, S.P., Brazil.

Weisz, P.B., and Marshall, J.F. 1979. High-grade fuels from biomass farming: potentials and constraints. Science 206:24-29.

World Bank. 1980. Alcohol Production from Biomass--Potential and Prospects in the Developing Countries. World Bank, Washington, D.C., USA.